D0803187

WESTERN ASTROLABES

Historic Scientific Instruments of the Adler Planetarium & Astronomy Museum

VOLUME I

Western

Bruce Chandler
General Editor

Sara Schechner Genuth
Editor

Astrolabes

by Roderick and
Marjorie Webster

with an
Introduction by
Sara Schechner
Genuth

ISBN 1-891220-01-2

All photographs are by Steven Pitkin, Ron Testa, or R. S. Webster, except as noted.
Copy Editing: Peggy Liversidge
Design: Pamela Geismar
Composition: Blue Friday Typographics
Printing: Toppan Printing Company

Printed in Japan

Adler Planetarium & Astronomy Museum
1300 South Lake Shore Drive
Chicago, Illinois 60605

Table of Contents

vii Foreword
ix History of the Adler Planetarium Collection
xiii Acknowledgments

INTRODUCTION

2 Astrolabes: A Cross-Cultural and Social Perspective

ASTROLABES

28 The Astrolabe: A Technical Introduction
40 Western Astrolabe Catalogue 1-35

ASTROLABE-QUADRANTS

126 The Astrolabe-Quadrant
130 Astrolabe-Quadrant Catalogue 36-45

MARINER'S ASTROLABES

148 Mariner's Astrolabe Catalogue 46-47

APPENDICES

152 Star Catalogue
160 Comparison of Stars
161 Concordance
162 Makers' Biographies
164 Bibliography

177 Index

We dedicate this volume to our friend, Francis R. Maddison,
whose depth of knowledge and high standards of scholarship
have set an example for us to try to follow.

Foreword

by Paul Knappenberger

One might reasonably expect to find the scientific instruments of antiquity in the venerable institutions of such cities as Oxford, England, or Florence, Italy, but not think of searching for such instruments in Chicago. Yet the Adler Planetarium houses a collection of antique scientific instruments rivaling in scope and quality any such collection, including those in Oxford and Florence.

In 1933 Philip Fox, the first director of the Adler, wrote of its antique instrument collection: "The items are marvellous not only for the learning and mechanical skill they display but for the beauty of workmanship." These pieces, indeed, are not only accurate mechanical representations of the movements of objects in the celestial sphere or relics of the technologies of past times. They are also intricately and precisely crafted *objets d'art* of fine wood, brass, ivory, enamel, and other rich materials. Furthermore, the combination of science, art, and craftsmanship in each instrument embodies the crafter's own ideas, sometimes representing a world-view — whether the earth or the sun occupies the center of the known universe, for example. Thus, each artifact of the Adler Collection can be considered an icon for the social order, beliefs, and customs of the time of its creation.

The Adler's "Universe In Your Hands" exhibition, which opened in the spring of 1995, was designed to reflect this interpretation of the Collection. This exhibition uses the artifacts to allow visitors to explore concepts, people, history, and science, including the practical applications of astronomy, through the ages. As a whole, the Collection is a finely detailed illustration of the timelessness of humanity's fascination with the cosmos.

The great size and variety of the Adler Collection — some 1,600 pieces ranging from armillary spheres to Zeiss star-plates — require a comprehensive catalogue in multiple volumes. The first two volumes of the Collection focus on one of the oldest-known types of scientific instruments, the astrolabe. Future catalogues will include two volumes on sundials and timekeeping devices, one on surveying and navigation instruments, and one on globes, telescopes, orreries, and other instruments.

The creation of the volumes of this catalogue is involving the efforts of many scholars worldwide, to whom I am grateful. It is also a pleasure to acknowledge the generous financial support for this project provided to date by the National Endowment for the Humanities, several individuals, and the Trustees of the Adler Planetarium.

Finally, I want to express a debt of gratitude to the Planetarium's *Curators Emeriti,* Roderick and Marjorie Webster, who have devoted so much of their own time, funds, and loving care to the Collection. In curatorial service since 1962, the Websters not only conserved the original instrument collection the Planetarium received from Max Adler in the 1930s, they also expanded the Collection through careful accessioning of rare pieces. The Websters' interest in astronomical artifacts began some 40 years ago with the purchase of a small scientific instrument — a pocket sundial. Their own search to discover more about antique instruments took them to the British Museum — and the experts there guided the Websters back to Chicago, not far from their own home. After many years' study of antique instruments, the Websters are now experts in their own right, and as much a treasure of the Adler as the instruments themselves.

It is my hope that this volume and the ones that follow will serve as guidebooks for others in search of discovery.

History of the Adler Planetarium Collection

Max Adler c. 1930

Chicago owes much to its early business leaders for their generous gifts to the city; among them are Marshall Field, who gave the Field Museum; John G. Shedd, who gave the Shedd Aquarium; Julius Rosenwald, who gave the Museum of Science and Industry; and Max Adler, Rosenwald's brother-in-law and the executive vice-president of Sears Roebuck and Co., who gave the Adler Planetarium.

In 1930, when the Planetarium opened to the public, its facilities included an astronomical museum. The main source of the museum's artifacts at that time was the Mensing Collection of early scientific instruments, around 550 items in 498 lots that Max Adler, in 1929, had been persuaded to purchase by Dr. Philip Fox, the head of the Astronomy Department at Northwestern University and the Planetarium's first director.

Anton W. M. Mensing, the managing director of the Müller Auction House in Amsterdam from the beginning of the twentieth century, was a friend of Raoul Heilbronner, a German antique dealer who lived and worked in Paris. Heilbronner had a personal collection of early measuring instruments of various kinds and was active in that market early in the 1890s, as shown by various Puttick and Simpson London auction sales during that era, when he was both a buyer and a seller of instruments.

Just before World War I was declared, Heilbronner returned to Germany in some haste, leaving his stock and personal collection behind. The French government sequestered all of his holdings and disposed of them through a series of auction sales once the war was over. The final sale, comprising scientific instruments, maps, globes, and books, had no catalogue, just a four-page brochure with two plates illustrating 27 instruments, most of which are at the Adler today. These artifacts were offered by the French government with a reserve of 60,000 francs on March 8, 1922.

The Heilbronner Sale Announcement, 1922

Many years ago Henri Michel of Brussels gave us his copy of that brochure. In an unknown hand was written "Vendue 240,000 Fs + 17%"; underneath this, Michel had written "Vendu à Mensing A'dam" and, some years later, "Vendu en 1932 au musée de Chicago 70,000 g."[1] Mensing bought the whole collection and incorporated his own private collection of instruments with Heilbronner's.

In 1924 Mensing had Dr. Max Engelmann, the curator at the Mathematisch-Physikalisch Salon in Dresden, write a catalogue of the combined collection. Since the Engelmann catalogue stops at No. 438, we thought for a long time that lots No. 439 to 498 had belonged to Mensing before he bought the Heilbronner collection, but in one of the letters in the Adler Planetarium's archives,[2] Mensing told Fox that Engelmann had not finished listing all of the instruments due to lack of time. We also know that between 1922 and 1924, Mensing disposed of some of the combined collection, as a few of the makers mentioned in the brochure and in a typed notice that he had sent out do not appear in the Engelmann catalogue.

Perhaps Mensing had bought the collection with the idea that Heilbronner, who had survived the war, could make a comeback as a dealer, but this did not happen, and Mensing ended up selling the items to at least three museums. Max Adler bought the instruments for the Planetarium in 1929, while the globes, maps, and books were divided between the National Maritime Museum at Greenwich (through the generosity of Sir James Caird)[3] and the Nederlands Scheepvaart-museum in Amsterdam.[4]

Sammlung Mensing, *Engelmann (1924)*

As none of Heilbronner's records survived, there was little history attached to his collection. Instruments known to have belonged to Heilbronner are shown in the two plates mentioned earlier. In addition, he had loaned fifteen instruments to the Exposition Universelle Internationale in Paris in 1900.[5] Among them we can recognize M-1, an armillary sphere by Gualterus Arsenius; M-36, an

1. Adler Planetarium Archives, under "Mensing files."

2. *Ibid.*

3. Adler Planetarium Archives, under "Mensing Collection, related material."

4. Communication from Willem Mörzer Bruyns, Nederlands Scheepvaartmuseum.

5. Paris Universal Exhibition Catalogue (1900), 22, 43–44.

View of the Mensing Collection, Amsterdam, 1924

astrolabe made in Toledo by Muhammed ibn Yusuf; and M-328, a marble polyhedral sundial. The descriptions are a bit sketchy, *e.g.*, "Grand astrolabe persan, en cuivre gravé" and "Petit astrolabe persan."

We also know that M-441, a Blaeu celestial globe, was originally part of Mensing's collection, not Heilbronner's, as Mensing mentioned in a letter to Fox that he had bought it and claimed it came from Rubens' house in Antwerp. Besides being a highly successful salesman, Mensing also had a workshop that repaired or restored missing parts to some of the instruments.

Between 1933 and 1958, the Adler Planetarium's collection of scientific instruments grew as the result of additional purchases made possible by the Adler Planetarium Trust, which had been set up by Max Adler from the Planetarium's share of the admission charges from the 1933-34 Chicago World's Fair. Other instruments, as well as books and ephemera, were given or purchased over the years.

In the Planetarium's cataloguing system, numbers beginning with "M" stand for Mensing. The early "A"s are from the Adler Trust, while the later ones represent additions from other sources. "G"s are for gifts, a classification that is no longer in use, while "L"s are for loans, many of which have come from the U.S. Navy.

The John Tomlinson Collection of portable sundials, designated "T," was purchased in 1940 from John Tomlinson, Jr.'s widow in New York. It included a number of sundials made by D. B. Sheahan for his and his friends' pleasure. They were signed with a number of fanciful names, such as "A. Dürer" and "T. J., Virginia." Mr. Sheahan, a maker of monumental and wall sundials that can be found on many campuses along the East Coast, became enamored of pocket sundials after reading Dr. Lewis Evans' chapter on portable sundials in the fourth edition of Mrs. Gatty's *Book of Sundials*. In the same collection, however, there

The Mensing Astrolabes, Amsterdam, 1924

were some very important original instruments, including T-5, a compendium from the workshop of Christopher Schissler (d. 1609), which formed part of the Mme. van der Does de Willibois sale in 1911 (the so-called Strozzi Sale). There was also T-35 (cat. no. 39), a special type of quadrant called a Panorganon, by Walter Hayes (fl. 1642-1692) of London; only four other examples of this type are known to us.

Kenneth Nebenzahl, a long-time trustee and a past chairman of the Board of Trustees of the Adler, gave 71 interesting instruments (designated "N") in 1985, and the Websters, as curators and trustees, have contributed their collection of instruments and early books (designated "W").

In 1986, David Pingree Wheatland donated 56 instruments, two mezzotints by Penther after Joseph Wright of Derby's paintings *The Orrery* and *The Air Pump,* and other ephemera. The instruments have been given the prefix "DPW." The most recent addition is the late Stillman Drake's collection of sectors (designated "SD"), including one that was probably made in Galileo's workshop.

The Adler Planetarium's recently formed History of Astronomy Department holds all of the early scientific instruments and has two libraries, one of modern reference works and the other of early books on instruments and the history of astronomy. The purchase of Saul Moskowitz's reference library in 1989 added material to both libraries. Other holdings include a large number of photographs, 3,000 of which were a gift from the late Dr. Derek J. de Solla Price of Yale University, as well as many early broadsides, celestial maps, and other ephemera. To date, the instruments and early books each number around 1,500 items.

Acknowledgments

We must first acknowledge the late Professor Derek J. de Solla Price, Yale University, for all his help and encouragement for our projects and for giving us many publications and photographs that have aided our research.

To Clare Vincent, Department of European Sculpture and Decorative Arts, Metropolitan Museum of Art, Professor Bernard Goldstein, University of Pittsburgh, and Professor David A. King, Frankfurt University, each of whom contributed one or more sections to specific entries, we extend our thanks for their scholarship and time.

We thank Steve Lubar, National Museum of American History, for his preparation of the initial star lists for most of the entries, during that long-ago summer he spent at the Adler.

We thank Andrea Murschel, Adler Planetarium, for her patience and capabilities. She coordinated David Pingree's work and developed the references and the Eastern bibliographical entries. She was also most helpful in producing the necessary photographs.

We also thank Paul Knappenberger, President of the Adler Planetarium, for his encouragement and support for the whole catalogue project.

We thank Sara Schechner Genuth, former curator at the Adler Planetarium and now at the University of Maryland, for her contributions to this catalogue. Her introduction is the keystone of this catalogue.

Without the abilities of Pamela Geismar, our designer, Lucy Sisman, our consultant, and Peggy Liversidge, our copy editor, this catalogue would never have made it to print, and we thank them heartily.

We also wish to thank Professor Owen Gingerich, Harvard-Smithsonian Center for Astrophysics, John Lamprey, Jonathan Snellenburg, and Anthony Turner for their help and cooperation.

Last, but not least, we owe a big debt of gratitude and thanks to our stalwart friend, Professor Bruce Chandler, the College of Staten Island, CUNY, our loyal, patient, hard-working editor, without whom all of our labors would have come to naught.

Introduction

Astrolabes: A Cross-Cultural and Social Perspective

Sara Schechner Genuth

"Little Lewis," Chaucer told his ten-year-old son as he packed him off to Oxford, "I have perceived well…thy ability to learn sciences touching numbers and proportions; and I [have] also consider[ed] thy earnest prayer specially to learn the Treatise of the Astrolabe." Here is "an astrolabe for our horizon," he continued, and a "little treatise…to teach thee a certain number of conclusions appertaining to the same instrument."[1] Chaucer's instrument — a planispheric astrolabe — was a portable model of the heavens, simulating the apparent rotation of the stars around the North Celestial Pole. It was also an analogue computer, which could be used to solve astrological and astronomical problems.[2]

Although Chaucer subtitled his treatise "Bread and Milk for Children," the astrolabe had been in the hands of scholars since antiquity. Indeed, an astronomer was seldom depicted without an astrolabe, the instrument becoming a badge of the profession.[3] (Figures 1-3) By the Renaissance, the astrolabe had also earned the respect of those whose work was not strictly astronomical. In the words of Jacob Köbel, this "marvelously delightful" instrument was "not only very practical and even necessary for astrologers, doctors, geographers, and others cultivating the arts and sciences, but also truly advantageous for mechanics [and] certain artisans."[4]

ORIGIN AND DIFFUSION

The origin of the astrolabe is somewhat obscure. The stereographic projection, which lies at the heart of its construction, was likely known by Hipparchus (*c.* 150 B.C.) and familiar to readers of *De architectura,* for in that work, Vitruvius (d. post-A.D. 27) described an anaphoric clock that used the projection on its dials. Ptolemy treated the projection in a theoretical way in his *Planisphaerium* (*c.* A.D. 160) and referred to a horoscopic instrument resembling an astrolabe at the end of this work. Even though Ptolemy's instrument consisted of a starry rete that turned on top of a plate marked with altazimuth coordinates, it likely lacked an

1. Geoffrey Chaucer, *Treatise on the Astrolabe* (1391); revised/modernized text based on the Rawlinson MS printed in Robert T. Gunther, *Chaucer and Messahalla on the Astrolabe*, Early Science in Oxford, vol. 5 (Oxford, 1929a), 1; *cf.* Geoffrey Chaucer, *A Treatise on the Astrolabe addressed to his son Lowys, A.D. 1391*, ed. Walter W. Skeat (London, 1872); and *idem*, *The Riverside Chaucer*, ed. Larry D. Benson, 3rd ed. (Boston, 1987), 661-83. Chaucer's treatise was based on an enormously popular Latin text, which survives in nearly 200 manuscripts and which was long ascribed to Māshā'allāh, but is now known to be a 13th-century, Latin compilation. See Paul Kunitzsch, "On the Authenticity of the Treatise on the Composition and Use of the Astrolabe Ascribed to Messahalla," *Archives internationales d'histoire des sciences* 31 (1981b): 42-62. For more on Chaucer's interest in astronomy, see John D. North, "Kalenderes Enlumyned Ben They: Some Astronomical Themes in Chaucer," *The Review of English Studies*, n.s., 20 (1969): 129-54, 257-83, 418-44; and *idem*, *Chaucer's Universe* (Oxford, 1988).

2. General works on the astrolabe include Robert T. Gunther, *The Astrolabes of the World*, 2 vols. (Oxford, 1932); Willy Hartner, "The Principle and Use of the Astrolabe," in *A Survey of Persian Art*, ed. Arthur Upham Pope, 3: 2530-54 (London/New York, 1939); *idem*, "Asṭurlāb," in *Encyclopedia of Islam*, new ed., 1: 722-28 (1960); both reprinted in *idem*, *Oriens-Occidens*, 2 vols. (Hildesheim, 1968/1984), 1: 287-318; Henri Michel, *Traité de l'astrolabe* (Paris, 1947); Leo Ary Mayer, *Islamic Astrolabists and Their Works* (Geneva, 1956); John D. North, "The Astrolabe," *Scientific American* 230 (1974): 96-106; reprinted in *idem*, *Stars, Minds and Fate: Essays in Ancient and Medieval Cosmology* (London, 1989), 211-20; National Maritime Museum, *The Planispheric Astrolabe* (Greenwich, 1976); Sharon Gibbs with George Saliba, *Planispheric Astrolabes from the National Museum of American History* (Washington, D.C., 1984); Roderick S. Webster, *The Astrolabe: Some Notes on Its History, Construction and Use*, 2nd ed. (Lake Bluff, Ill., 1984); A. J. Turner, *Astrolabes, Astrolabe Related Instruments*, The Time Museum: Catalogue of the Collection, ed. Bruce Chandler, vol. 1: Time Measuring Instruments, part 1 (Rockford, Ill., 1985); Owen Gingerich, "Zoomorphic Astrolabes and the Introduction of Arabic Star Names into

alidade. Ptolemy's planisphere performed the functions of a star-finder, but it appears not to have been used for directly measuring the altitudes of stars. The instrument we have come to know as the astrolabe — complete with rete, tympans, and alidade — was invented sometime before the late fourth century, when Theon of Alexandria wrote a tract on it. Treatises by John Philoponus (A.D. 530), Severus Sebokht (pre-660), al-Fazārī (late 8th century), and others followed in Greek, Syriac, Arabic, and lastly Latin, as knowledge of the instrument was diffused. Early manufacture was centered on Ḥarrān, a city between the Tigris and Euphrates rivers, which was also a hub for the translation of Greek and Syriac works into Arabic.[5]

Prior to the tenth century, knowledge spread eastward from the Syro-Egyptian region through Ḥarrān to Iraq and Persia.[6] In the *Mashriq* (the East), the astrolabe-making craft was highly esteemed during the reigns of the first ʿAbbāsid Caliphs, and that of al-Ma'mūn (787-827) in particular. The profession often passed down from father to son through several generations.[7] From Persia, interest passed to Mughal India in the mid-sixteenth century in the wake of new rulers, who regulated their affairs according to the principles of astrology and valued astrolabes for this purpose. Lahore (in what is now Pakistan) became a center for the production of Indo-Persian astrolabes.[8] South of Lahore, Hindus had been introduced to the astrolabe perhaps as early as the eleventh century by traveling scholars, such as al-Bīrūnī.[9] The first Sanskrit treatise on the astrolabe was composed around 1370. Few Indian astrolabes, however, predate the seventeenth century, and most surviving examples date from the eighteenth century onward,

FIGURE 1 *Astronomia with an astrolabe. Detail from Joannes de Sacrobosco,* Textus de sphera *(Paris, 1500). Courtesy of the Adler Planetarium, Chicago.*

Europe," in *From Deferent to Equant: A Volume of Studies in the History of Science in the Ancient and Medieval Near East in Honor of E. S. Kennedy*, ed. David A. King and George Saliba, 89-104, Annals of the New York Academy of Sciences, vol. 500 (New York, 1987); David A. King, *Islamic Astronomical Instruments* (London, 1987a); and *idem*, "Die Astrolabiensammlung des Germanischen Nationalmuseums," trans. Kurt Maier, in Germanisches National Museum, *Focus Behaim Globus*, exhibition catalogue edited by Gerhard Bott, 2 vols., 1: 101-14, 2: 568-602, 640-43 (Nuremberg, 1992).

3. On astronomers and their obligatory astrolabes, see, for example, an illuminated Hebrew manuscript from Spain, *c.* 1350-1399 (Copenhagen, Det Kongelige Bibliotek, Cod. Hebr. 37, fol. 114r); another Hebrew-Spanish manuscript, 1472 (Oxford, Bodleian Library, MS Kennicott 1, fol. 90r); and a Hebrew manuscript from Germany, *c.* 1400-1450 (London, British Library, MS Or. 10878, fol. 17r); all are discussed in Thérèse Metzger and Mendel Metzger, *Jewish Life in the Middle Ages: Illuminated Hebrew Manuscripts of the Thirteenth to the Sixteenth Centuries* (New York, 1982), 157, 166. The Spanish manuscript in Oxford depicts Balaam, who was identified as an astronomer in Jewish tradition and was shown holding an astrolabe in some Biblical miniatures. For examples from the Christian tradition, see a Czech manuscript of *Mandeville's Travels* (London, British Library, Add. MS 24189, fol. 15r); reproduced in Josef Krása, ed., *The Travels of Sir John Mandeville: A Manuscript in the British Library*, trans. Peter Kussi (New York, 1983), plate 19; Hartmann Schedel, *Büch der Cronicken* (Nuremberg, 1493), fol. CCLVr, which shows Regiomontanus with an astrolabe; and Johannes Kepler, *Tabulae Rudolphinae* (Ulm, 1627), frontispiece. Astronomia with an astrolabe appears in a woodcut in Joannes de Sacrobosco, *Textus de sphera* (Paris, 1500). An astronomer was often depicted with an astrolabe, just as a physician was typically shown with a flask of urine; see Schedel (1493), *passim*.

4. Jacob Köbel, *Astrolabii declaratio, eiusdemque usus mire jucundus, non modo astrologis, medicis, geographis, caeterisque literarum cultoribus multum utilis ac necessarius; verum etiam mechanicis quibusdam opificib. non parum commodus* (Mainz, 1535).

5. Otto Neugebauer, "The Early History of the Astrolabe," *Isis* 40 (1949): 240-56; *idem*, *A History of Ancient Mathematical Astronomy*, 3 vols. (New York, 1975), 2: 868-79; Paul Kunitzsch, "Observations on the Arabic Reception of the Astrolabe," *Archives internationales d'histoire des sciences* 31 (1981a): 243-52; David A. King, "The Origin of the Astrolabe According to Medieval Islamic Sources," *Journal for the History of Arabic Science* 5 (1981): 43-83; reprinted in *idem* (1987a); Hartner (1968/1984), 1: 288-90; Turner, A. J. (1985), 10-14. *Cf.* the introduction of A. P. Segonds, in Jean Philopon, *Traité de l'astrolabe,* ed. and trans. A. P. Segonds, *Astrolabica* 2 (Paris, 1981).

6. Turner, A. J. (1985), 14, 21-22.

7. Hartner (1968/1984), 1: 289-90.

8. Sayyid Sulayman Nadvi, "Indian Astrolabe Makers," *Islamic Culture* 11 (1937): 537-39; Turner, A. J. (1985), 25-26.

9. Turner, A. J. (1985), 26.

FIGURE 2 *Regiomontanus with an astrolabe. From Hartmann Schedel,* Büch der Cronicken *(Nuremberg, 1493). Private collection.*

suggesting that interest in Sanskrit circles was awakened by the observatory-building program of the Maharajah Jai Singh II (1686-1743).[10]

Traveling to the northeast, the astrolabe reached China by 1267, when Jamāl al-Dīn, a Persian astronomer, brought Kublai Khān models of some astronomical instruments that equipped the Marāghah observatory newly erected by his brother, Hūlāgū Khān.[11] Within ten to twenty years, Marco Polo claimed to have observed the widespread use of a kind of astrolabe among astrologers in Beijing, and *The Travels of Sir John Mandeville,* written around 1357, described the instrument's importance to the philosophers at Kublai Khān's table.[12] There are reasons, however, to question the accuracy of these two European reports. First, no astrolabes (or texts describing them) are known to be preserved in China.[13] Second, it seems implausible for the astrolabe to have achieved the popularity Polo claimed for it within twenty years of Jamāl al-Dīn's visit, for the Chinese had no prior knowledge of stereographic projections, spherical trigonometry, and the twelve-partite zodiac, which underlay the design of the astrolabe.[14] Polo may have been mistaken in identifying the instruments he saw, or he may

10. David Pingree, "Islamic Astronomy in Sanskrit," *Journal for the History of Arabic Science* 2 (1978b): 315-30; *idem*, "History of Mathematical Astronomy in India," in *Dictionary of Scientific Biography*, 15: 533-633, esp. 626-28 (New York, 1978a); Virendra Nath Sharma, "The Great Astrolabe of Jaipur and Its Sister Unit," *Archaeoastronomy* no. 7, *Supplement to the Journal for the History of Astronomy* 15 (1984): s126-28; Turner, A. J. (1985), 26-28.

11. Willy Hartner, "The Astronomical Instruments of Cha-ma-lu-ting, Their Identification, and Their Relations to the Instruments of the Observatory of Marāgha," *Isis* 41 (1950): 184-94; Joseph Needham, *Science and Civilisation in China*, 6 vols. (Cambridge, England, 1954-88), 3: 372-74.

12. Henry Yule, trans. and ed., *The Book of Ser Marco Polo The Venetian Concerning the Kingdoms and Marvels of the East*, 3rd ed., rev. Henri Cordier, 2 vols. (London, 1903), 1: 446-47, containing book 2, chap. 33: "Concerning the Astrologers in the City of Cambaluc [Beijing]." Polo wrote that amongst the Christians, Saracens, and Cathaians, there were five thousand astrologers provided with annual maintenance and clothing by the Great Khan. "They have a kind of astrolabe on which are inscribed the planetary signs, the hours and critical points of the whole year." All three sects used their instruments for the practice of natural and judicial astrology and answered horary questions for many people. M. C. Seymour, ed., *Mandeville's Travels* [*c.* 1357] (London, 1968), chap. 25: 179; this reference is discussed below, on pp. 13-14.

13. Needham (1954-88), 3: 376.

14. Hartner (1950), 192; *cf.* Needham (1954-88), 3: 375-77.

FIGURE 3 *Hebrew astronomer holding an astrolabe and consulting astronomical tables. Miniature illustrating the second part of Maimonides'* Guide to the Perplexed, *in a manuscript executed in Barcelona in 1348 by Levi ben Isaac hijo Caro of Salamanca. By permission of Det Kongelige Bibliotek, Copenhagen, Cod. Hebr. 37, fol. 114*[r]*.*

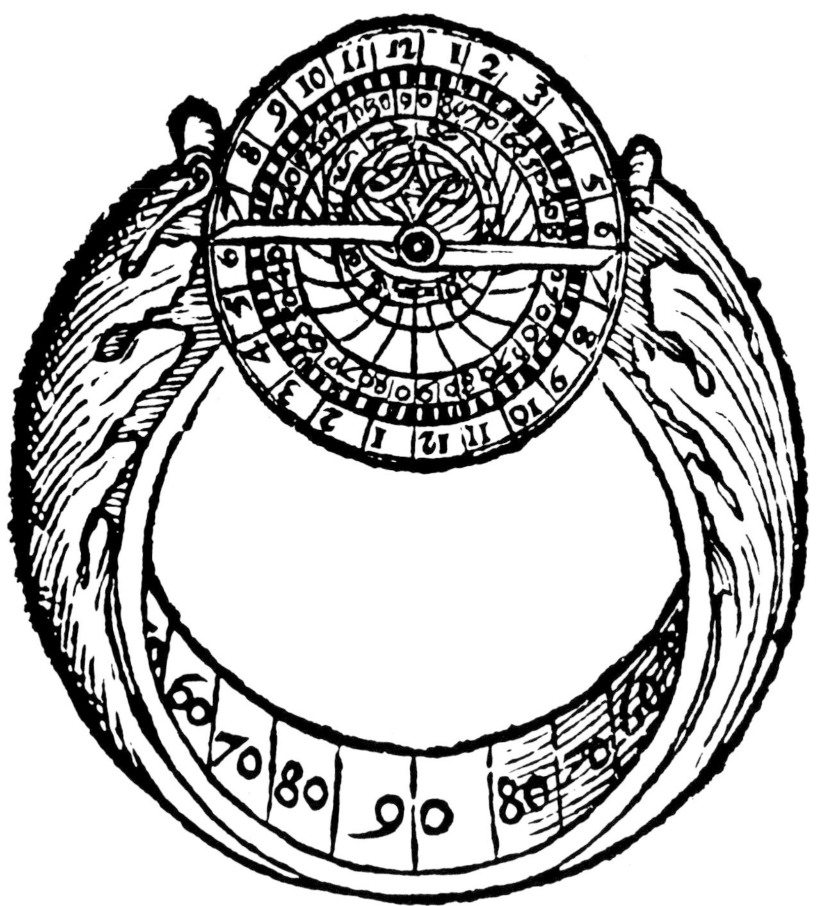

FIGURE 4 *Astrolabe ring of Bonet de Lattes. From* Annuli astronomici *(Paris, 1558). Courtesy of the Adler Planetarium, Chicago.*

have called them astrolabes for want of a better term.[15] In any case, it seems that the astrolabe did not catch on in China. If it found favor at all, it would appear to have been among foreigners. In this regard it is interesting to note that some medieval Persian astrolabes list China on their gazetteers.[16]

Traveling in the other direction, knowledge of the astrolabe spread westward to North Africa (the *Maghrib*) and Muslim Spain (Andalusia) by the tenth century, and from there to Christian Europe. Arabic texts were translated into Latin, and Jewish scholars contributed to the transmission process by translating into Hebrew.[17] Students from northern Europe crossed into Spain and returned with treasures of Greco-Arabic science. Prominent among them was Gerbert of Aurillac (*c.* 945-1003), who became Pope Sylvester II in 999. Gerbert introduced the astrolabe to his pupils at Rheims and elsewhere on his return from Catalonia.[18] Within 50 years, Hermann Contractus of Reichenau (1013-1054) prepared an adaptation of *De utilitatibus astrolabii,* one of the treatises likely imported by Gerbert.[19] And in 1092, Walcher (d. 1135), prior of the Abbey of Malvern in England, observed the time of a lunar eclipse with his instrument.[20] The diffusion process, apparently, was well under way. By the mid-thirteenth century, the astrolabe was known to scholars as a teaching, calculating, and observing instrument. Astrolabes were

15. Another way to reconcile this is to discount his claim that the astrolabe was used by Christians, Saracens, and Cathaians alike and assume that it was principally used by the Christians and Muslims to whom he refers.

16. See astrolabes made by ibn Dawlatshah, 1388 (N-70) and ibn Ja'far, 1433/4 (A-91). See the A. P. catalogue of Eastern astrolabes (forthcoming).

17. J. M. Millàs-Vallicrosa, "Translations of Oriental Scientific Works (to the End of the Thirteenth Century)," in *The Evolution of Science*, ed. Guy S. Metraux and François Crouzet, 128-67 (New York, 1963); David C. Lindberg, "The Transmission of Greek and Arabic Learning to the West," in *Science in the Middle Ages*, ed. David C. Lindberg, 52-90 (Chicago, 1978); *Encyclopaedia Judaica, s.v.* "Astrolabe."

18. See Millàs-Vallicrosa (1963), 139-43, on Gerbert's sojourn in Spain and his contact with the treatises on the astrolabe prepared at the Monastery of Santa Maria de Ripoll.

19. *Ibid.*, 146; Lindberg (1978), 60-61; *cf.* Mary Catherine Welborn, "Lotharingia as a Center of Arabic and Scientific Influence in the Eleventh Century," *Isis* 16 (1931): 188-99.

20. Charles Homer Haskins, *Studies in the History of Medieval Science*, 2nd ed. (Cambridge, Mass., 1927), 114-15; *cf.* Emmanuel Poulle, *Walcher de Malvern et son astrolabe (1092)*, Centro de Estudos de Cartografia Antiga, no. 132 (Coimbra: Junta de Investigações Científicas do Ultramar, 1980b).

21. For lectures on the astrolabe at Bologna, see Lynn Thorndike, *University Records and Life in the Middle Ages* (New York, 1944), 281, 403. Astronomical codices typically contained Sacrobosco's *Tractatus de sphaera*, *Compotus*, and *Algorithmus*, the *Theorica planetarum*, astronomical tables, and treatises on the astrolabe and quadrant; see Edward Grant, ed., *A Source Book in Medieval Science* (Cambridge, Mass., 1974), 451; and Olaf Pedersen, "Astronomy," in *Science in the Middle Ages*, ed. David C. Lindberg, 303-37 (Chicago, 1978). Medieval treatises and astrolabe observations are discussed in Haskins (1927), *passim.* On borrowing astrolabes, see Turner, A. J. (1985), 30 n. 99.

22. Emmanuel Poulle, "L'astrolabe médiévale d'après les manuscrits de la Bibliothèque Nationale," *Bibliothèque de l'École de Chartes* 112 (1954): 81-103, esp. 89. In the Duomo of Pisa, one can still see the figure of a muse wielding an astrolabe, which Giovanni Pisano carved into the Gothic pulpit he created between 1302 and 1311.

23. A. J. Turner (1985), 26, suggests that the presence of few Indo-Persian astrolabists in Lahore might indicate that astrolabes were not in general use throughout the society, but were reserved for members of the Mughal court.

24. Lynn Thorndike, *Michael Scot* (London, 1965), 32.

25. Metzger and Metzger (1982), 157.

26. All in all, Charles V's collection included seven astrolabes of copper, two of brass, two of silver, and one of gold. See Jules Labarte, *Inventaire du mobilier de Charles V, roi de France* (Paris, 1879), line items 1990, 2072, 2216, 2270 (2 astrolabes), 2427 (3 astrolabes), 2714, 2817, 3119, 3121; and Turner, A. J. (1985), 32, 34.

27. George H. Gabb, "The Astrological Astrolabe of Queen Elizabeth," *Archaeologia* 86 (1937): 101-3; R. T. Gunther, "The Astrolabe of Queen Elizabeth," *Archaeologia* 86 (1937b): 65-72.

28. Bonet de Lattes, *Annulus astronomicus* (Rome, 1493); reprinted many times as *Annuli astronomici utilitatum liber*. See, for example, Johann Dryander, *Annulorum trium diversi generis instrumentorum astronomicorum, componendi ratio atq; usus, cum quibusdam aliis lectu iucundissimis* (Marburg, 1537), sigs. Hiij^r-Kiij^r; or *Annuli astronomici, instrumenti cum certissimi, tùm commodissimi, usus, ex variis authoribus, Petro Beausardo, Gemma Frisio, Ioãne Dryandro, Boneto Hebraeo, Burchardo Mythobio, Orontio Finaeo* (Paris, 1558), fols. 103^v-117^v. It is interesting to note that the illustration of the ring in Dryander (1537) is surrounded by the words "ANNULUS BONETI / INEVITABILE FATUM" (once repeated), suggesting Bonet's ring revealed one's inescapable fate. At first glance, this seems to contradict my claim that wearing an astrolabe was empowering, but this is not so. To be forewarned was to be forearmed, and astrology was valued because it offered the prospect of control over future effects and repercussions. For more on Bonet de Lattes, see *Encyclopaedia Judaica*, *s.v.* "Lattes, Bonet."

frequently found with treatises on their use in university libraries; they could even be borrowed.[21]

Evidence of cultural diffusion appeared as early as the twelfth century, when Héloise and Abelard named their son "Astrolabe" and images of the instrument began to appear in cathedral sculpture and miniatures.[22] Beyond the ken of universities and cathedrals, astrolabes made their way into court, where court astrologers in the West found them as indispensable as those in the East.[23] When traveling with Frederick II (1194-1250), Michael Scot discussed astronomical matters with the Holy Roman Emperor and took readings from his astrolabe (perhaps to predict the outcome of military campaigns).[24] A Jewish professor of astronomy and astrology at the University of Salamanca, Abraham Zacuto, similarly served John II and Manuel I of Portugal in the fifteenth century.[25] Charles V (1337-1380) of France deemed the astrolabe so essential that he owned at least twelve, including one of gold and two of silver.[26] Elizabeth I (1533-1603) of England had two of gilt brass, one of which was dated 1559, the year of her coronation. Elizabeth had come to power on an astrologically propitious day (selected by John Dee), and the symbolism of her coat-of-arms being surrounded by tables of planetary virtues on the back of one astrolabe was not likely lost on the queen.[27] A similar touch appeared in an astrolabe ring invented by Bonet de Lattes (d. *c.* 1514), a rabbi who became doctor and astrologer to Pope Alexander VI in 1498 and later served Pope Leo X. In wearing this ring, a commanding individual might believe he had the power of the heavens close at hand.[28] (Figure 4)

The apogee of astrolabe production in Europe occurred during the late fifteenth and sixteenth centuries, and many tracts tried to help

FIGURE 5 *Astronomical compendium with astrolabic projections, by Christopher Schissler, Augsburg, 1559. On the left, a de Rojas projection of the celestial sphere. On the right, a north-polar, stereographic projection of the earth. Courtesy of the Adler Planetarium, Chicago (M-365).*

the novice learn his way around the instrument.[29] Notable makers included Jean Fusoris (*c.* 1365-1436) of Paris, whose career came to an end with his arrest on espionage charges; Gualterus Arsenius (fl. 1554-1579) of Louvain, who incorporated many innovations made by his uncle Gemma Frisius (1508-1555); and Georg Hartmann (1489-1564) of Nuremberg.[30] Hartmann's output was considerable, and in his workshop, craftsmen finished several astrolabes simultaneously. The evidence for a division of labor is to be found in the use of assembly marks. In Hartmann's case, these were numbers stamped in unobtrusive places on rough-hewn components that had been assembled. The parts were then separated, finished by different artisans, and reassembled into a complete instrument.[31]

In general, astrolabes were elegant instruments produced for elite patrons. In the East, Persian instruments were kept in small sacks and regarded by the common people as precious gems.[32] In the West, they were prized as "mathematical jewels."[33] They graced the cabinets of nobles and even ornamented the bindings of books, like those in the magnificent library of Jean, Duc de Berry (1340-1416), a son, brother, and uncle of three kings of France.[34] During the Renaissance, astrolabic devices were incorporated in astronomical compendia — the pocket-sized, gilt-and-silvered brass instruments that typically contained at least one sundial, a nocturnal, a magnetic compass, a gazetteer, a lunar volvelle, and sometimes an array of maps, tables, and scales.[35] Astrolabes also adorned fabulous clocks, which marked the rotation of the stars as well as the hours.[36] (Figures 5-7) Both compendia and astrolabe clocks were the sumptuous property only of the very wealthy. Like fine astrolabes, they were purchased by aristocrats not so much for daily use, but as a way to patronize the arts, give evidence of their erudition, and enhance their prestige. That is not to say that no one of wealth ever used an astrolabe for its intended astronomical purpose, but there is no doubt that many astrolabes have survived because they were showpieces preserved in the cabinets of patricians.

FIGURE 6 *Astrolabe clock, by Johann Christoff Lang, Augsburg,* c. *1580. Courtesy of the Adler Planetarium, Chicago (M-383).*

At the other end of the social scale, we find the do-it-yourself astrolabes. Astrolabes were sometimes produced as manuscripts, painstakingly inked on vellum or paper, and glued to wood or pasteboard.[37] These were labors of love. But for those with less patience and skill, Georg Hartmann pioneered the cheap, printed astrolabe, whose paper parts could be cut out and pasted to wood. Four of these instruments, dating from 1531 to 1540, survive. Also connected to the paper-astrolabe trade were Johannes Krabbe, Egnatio Danti, Willem Janszoon Blaeu, Philippe Danfrie, Jean Moreau, Henry Sutton, John Prujean, and Nicolas Bion.[38]

29. Some of the more important works include *Astrolabii quo primi mobilis motus deprehenduntur canones* (Venice, 1512); Johann Stöffler, *Elucidatio fabricae ususque astrolabii* (Oppenheim, 1513); Johann Copp, *Erklaerung unnd Gründtliche underweysung, alles nutzes, so in dem Edlen Instrument, Astrolabiū genañt* (Augsburg, 1525); Joannis Martini Poblacion [pseud. of Juan Martínez Siliceo], *De usu astrolabii compendium* ([Paris], 1527); Oronce Fine, *Quadrans astrolabicus* (Paris, 1534); Jacob Köbel, *Astrolabii declaratio* (Mainz, 1535); Jacques Focard, *Paraphrase de l'astrolabe* (Lyons, 1546); Juan de Rojas, *Commentariorum in astrolabium quod planisphærium vocant, libri sex* (Paris, 1550); Gemma Frisius, *De astrolabo catholico liber* (Antwerp, 1556); Dominique Jacquinot, *L'usage de l'astrolabe....plus est adjousté une amplification de l'usage de l'astrolabe, par Jacques Bassentin Escossois*, 2nd ed. (Paris, 1559); Egnatio Danti, *Trattato dell'uso e della fabbrica dell'astrolabio* (Florence, 1569); John Blagrave, *The Mathematical Jewel* (London, 1585); Robert Tanner, *The Traveller's joy and felicitie, or a Mirror for Mathematics* (London, 1587); and Christoph Clavius, *Astrolabium* (Rome, 1593). Many texts were reprinted, revised, or translated. See, for example, Jacques Focard, *Paraphrase de l'astrolabe*, rev. Jacques Bassentin (Lyons, 1555); Johann Stöffler, *Traité de la composition et fabrique de l'astrolabe, & de son usage....Le tout traduit du Latin de Iean Stofler de Iustingence.... Avecques annotations...faites par Iean Pierre de Mesmes* (Paris, 1560); and Gemma Frisius, *De astrolabo catholico liber* (Antwerp, 1583), reissued as part of Petrus Apianus, *Cosmographia* (Antwerp, 1584), 354-479.

30. The A. P. has instruments by each of these makers. For those by Fusoris, see M-27 (cat. no. 2) and W-264 (cat. no. 3); by Arsenius, M-23 (cat. no. 8) and M-24 (cat. no. 9); by Hartmann, M-22 (cat. no. 6) and W-272 (cat. no. 5). For brief portraits of these makers, see Emmanuel Poulle, *Un constructeur d'instruments astronomiques au XVe siècle: Jean Fusoris* (Paris, 1963); and Turner, A. J. (1985), 37-47.

31. One Hartmann astrolabe (M-22, cat. no. 6) in the A. P. Collection has the assembly number "2" stamped on its mater. On the production process in Persia, see John Chardin, *Voyages du Chevalier Chardin, en Perse, et autres lieux de l'Orient*, new ed., 4 vols. (Amsterdam, 1735), 3: 168-74.

32. *Ibid.*, 3: 168. The Persian astrolabe (A-40) of al-'Abd Amīn of Mashad, dated 1669/70 or 1688/9, has such a leather pouch. See the A. P. catalogue of Eastern astrolabes (forthcoming).

33. In his epistle dedicatory to Catherine de Medici, Jacquinot (1559), sig. ãij, alluded to the astrolabe and the mathematical sciences on which it was based as jewels fit for a queen. This tone had been set in France at least 150 years before Jacquinot, when astrolabes were literally treated as jewels by Jean, Duc de Berry (1340-1416), a passionate collector of rare gems and exquisite manuscripts. His library contained books not only fastened in gold and silver and studded with precious stones, but also encrusted with gilt astrolabes. On the Duc de Berry, see Jules Guiffrey, *Inventaires de Jean, duc de Berry (1401-1416)*, 2 vols. (Paris, 1894-96); and Jean Longnon and Raymond Cazelles, *The Très Riches Heures of Jean, Duke of Berry* (New York, 1969). In England, Blagrave (1585) called his new universal astrolabe a "mathematical jewel."

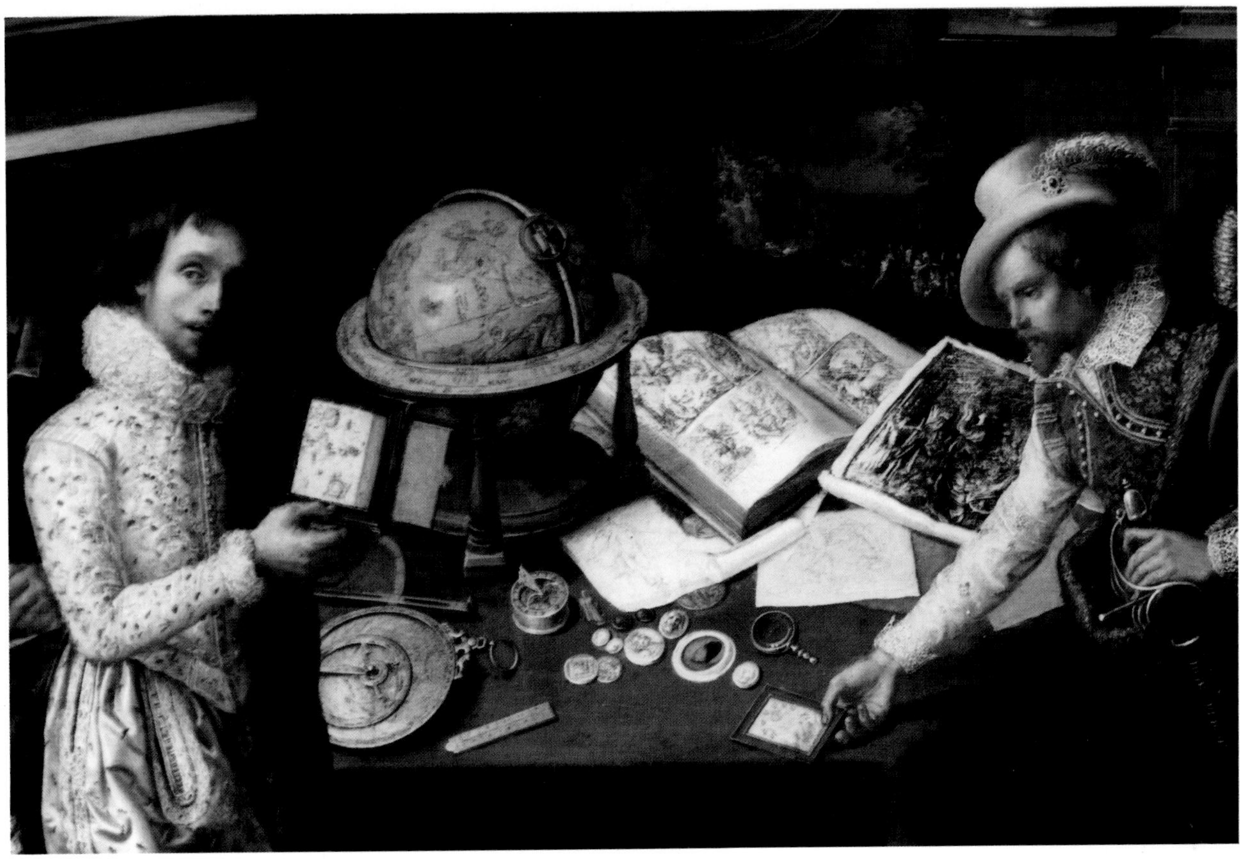

FIGURE 7 *Flemish astrolabe amidst other artifacts of art, science, and commerce. Detail from* Cognoscenti in a Room Hung with Pictures, *Flemish School, early seventeenth century. Courtesy of the National Gallery, London.*

Paper instruments may have helped to popularize the astrolabe — and there is evidence of their use by schoolboys — but they could not compensate for the deficiencies (by seventeenth-century standards) of the brass models.[39] On these small and portable instruments, the scales could not be finely divided, whereas larger and heavier astrolabes, which offered the prospect of being more precise, were inconvenient to hold. Astrolabes, moreover, were mathematically complex, and therefore difficult to construct and expensive to buy.[40] Finally, only those with a good grasp of geometry and trigonometry could learn to use them effectively.

By the seventeenth century, the astrolabe — the queen of medieval instruments — was dethroned in the West. For time-finding, the average person had simpler sundials and nocturnals at his or her disposal. For land measure, the surveyor turned to the new theodolite, graphometer, and circumferentor. For star-finding, the astronomer and astrologer consulted star charts and planispheres. And for demonstrating astronomical relationships and solving basic problems of positional astronomy, the tutor employed armillary spheres and globes. Many specialized instruments — some derived from the astrolabe — took the place of this very compact, multipurpose instrument.

Consequently, the production of astrolabes and companion textbooks in Europe slackened during the seventeenth century and was dead by the eighteenth. The instrument, nonetheless, continued to be a staple of Islamic workshops well into the nineteenth century. The

34. The Duc de Berry was the son of Jean II, le Bon (r. 1350-1364); the brother of Charles V (r. 1364-1380); and the uncle of Charles VI (r. 1380-1422); see note 33 above. On the contents of noble cabinets, see A. J. Turner, *Early Scientific Instruments: Europe, 1400-1800* (London, 1987), 57; and Oliver Impey and Arthur MacGregor, eds., *The Origins of Museums: The Cabinet of Curiosities in Sixteenth- and Seventeenth-Century Europe* (Oxford, 1985).

35. Astronomical compendia will be treated in the time-finding section of the comprehensive catalogue of the Adler collection. Noteworthy among the compendia are those dating from the 1550s, made by Christopher Schissler of Augsburg or his workshop (M-364, M-365, W-77, T-5); these contain stereographic or de Rojas projections.

36. Astrolabe clocks will be treated in the time-keeping section of the comprehensive catalogue of the Adler collection. Noteworthy among them are the Orpheus Clock (M-377), made in Germany, *c.* 1580, and a clock produced by Johann Christoff Lang, Augsburg, *c.* 1600 (M-383). On astrolabe clocks, see Catherine Cardinal, "Horloges de table astrolabiques françaises du XVIe siècle," *Astrolabica* 4 (1986): 3-20; and Klaus Maurice and Otto Mayr, eds., *The Clockwork Universe: German Clocks and Automata, 1550-1650* (New York, 1980).

37. See the Laurentius Schreckenfuchs astrolabe, dated 1567 (W-109, cat. no. 12), in this volume.

reason for its continued esteem had much to do with the different social role it played in the East.

USES OF THE ASTROLABE

As might be guessed of an instrument so widely dispersed and used over so many centuries, regional and cultural differences arose with respect to the instrument's design and use. Let us begin with the common features before turning to the variations.

The astrolabe's principal uses were astronomical, astrological, and topographical. For astronomical purposes, the astrolabe was described both as the "most perfect, certain, and necessary" and the "most pleasant, agreeable, and easy" instrument available to locate stars in the sky and discover the times of their risings and settings.[41] If the praise was a bit hyperbolic, it nonetheless conveyed the pride scholars had in owning so handsome and handy an instrument. With an astrolabe, an astronomer could find the altitude and azimuth, the ecliptic latitude and longitude, and the declination and ascension of any planet, comet, or star. He could find the parallax of the moon, the distance between stars, or the length of a comet's tail. He could also determine when a solar or lunar eclipse would occur, how long it would last, and how much of the celestial body would be darkened.[42] What's more, some of these calculations could be done without ever going outside, because the astrolabe was, in effect, an analogue computer. The instrument simulated the apparent rotation of the stars around the North Celestial Pole. Indeed, as John Blagrave remarked, an astrolabe could be used "to know the height of any starre above the horizon [when] sitting close within dores, and thereby to learne to know the starres in the skie."[43] Both Blagrave and Gemma Frisius extolled this pedagogical function.[44] The instrument could also be used indoors to calculate the length of day or night and to determine the beginning, end, and duration of twilight. Back outside, day or night, it could be employed to find the time.[45] Trigonometric scales for determining the sine or cosine of an angle were frequently inscribed on *mashriqi* astrolabes as well, making each example a mathematical tool as much as a star-finder or time-finder.

It has been claimed that the development of Latin astronomy in the eleventh and twelfth centuries owed much to the introduction of the astrolabe.[46] Although the arguments in support of this claim are rather tenuous, it appears that the counterclaim — that the astrolabe was little if ever used for observation and was employed solely as a calculating device or pedagogical model — also needs modification.[47] There is evidence that the astrolabe was occasionally used for astronomical research. In 1092, as we have seen, Walcher used his astrolabe to determine the time of a lunar eclipse, and in 1274, an unknown Danish astronomer at Roskilde measured the altitude of the sun each day in order to determine the length of daylight and compile a table for the calendar of the cathedral chapter.[48] In the early fourteenth century, moreover, Levi ben Gerson (1288-1344) took pains to establish astrolabic observations on a sure footing in his treatise on astronomy.[49] The astrolabe can also be seen in depictions of early observatories, such as that of Taqī al-Dīn in Istanbul, around 1577 (Figure 8); and Tycho Brahe

38. See, for example, the Danfrie astrolabe, 1584, reissued by Moreau in 1622 (W-98, cat. no. 19), in this volume. On makers of paper astrolabes or examples of their work, see Gunther (1932), 358-59, 401-2, 438-40, 448-50, 519; National Maritime Museum (1976), 37, 41; Turner, A. J. (1985), 43-44, 50; A. J. Turner, "Paper, Print, and Mathematics: Philippe Danfrie and the Making of Mathematical Instruments in Late 16th Century Paris," in *Studies in the History of Scientific Instruments*, ed. Christine Blondel, *et al.*, 22-42 (London/Paris, 1989); and King (1992), 2: 602-3. On paper instruments in general, see A. J. Turner, *Paper and Brass: Scientific Instruments and the Art of Printing. A Catalogue of an Exhibition held June...1974* (London, 1974); and Emmanuel Poulle, *Les instruments de la théorie des planètes selon Ptolémée: Équatoires et horlogerie planétaire du XIIIe au XVIe siècle*, 2 vols. (Geneva, 1980a), *passim*.

39. Oxford, Bodleian Library, MS Aubrey 10, fol. 109; quoted in A. J. Turner, "Mathematical Instruments and the Education of Gentlemen," *Annals of Science* 30 (1973): 51-88 (quotation on 65). For John Aubrey's comments (that every schoolboy should have a paper astrolabe), see below, p. 13.

40. Moreau notes these problems as justification for issuing paper instruments. See Jean Moreau, *L'usage de l'un et l'autre astrolabe particulier et universel* (Paris, 1625); sig. aII; quoted in Turner, A. J. (1989), 33.

41. Jacquinot (1559), sig. āij^v (my translation); and Denis Henrion, *Briefve explication de l'usage de l'astrolabe* (Paris, 1620), 2 (my translation); separately paginated, but included in Denis Henrion, *Collection, ou recueil de divers traictez mathematiques* (Paris, 1621).

42. See Gemma Frisius (1556), chaps. 47, 65-71, 74-76; and Blagrave (1585), 40-41, 50-61, 73.

43. Blagrave (1585), 38.

44. Gemma Frisius (1556), chap. 38; Blagrave (1585), 38-39.

45. On using the astrolabe to find the time, see North (1989), 218-19; *cf.* John D. North, "Astrolabes and the Hour-Line Ritual," *Journal for the History of Arabic Science* 5 (1981): 113-14.

46. Pedersen (1978), 309-14, reasons as follows: When Walcher used his astrolabe to determine the time of a lunar eclipse in 1092, he recognized that his observation could establish an epoch for a table of lunations, which he prepared in 1108 (the earliest table known to have been created by a Latin astronomer). In 1126, Adelard of Bath (fl. 1116-1142) translated the astronomical tables of al-Khwārizmī, and others drew on the work of al-Zarqālī and al-Battānī. With information on the mean motions of the sun, moon, and planets, these tables allowed astronomers to utilize their astrolabes more fully. The problem was that Western astronomers little understood the astronomical theories on which the tables were based. This impasse led to the translation of Greek and Arabic manuals of astronomy and planetary theory — culminating in the translation of Ptolemy's *Almagest* in the third quarter of the 12th century — and opened the door to new research using precision instruments.

47. Poulle (1980b), 3-4, takes this position.

48. Pedersen (1978), 312, 322.

49. Bernard R. Goldstein, "Levi ben Gerson: On Instrumental Errors and the Transversal Scale," *Journal for the History of Astronomy* 8 (1977): 102-12.

FIGURE 8 *Taqī al-Dīn's observatory, Istanbul,* c. *1577. Miniature from the* History of the King of Kings, *a poem by 'Alā ad-Dīn Mansūr-Shīrazī. Courtesy of Istanbul Üniversitesi Kutüphanesi, MS Yildiz 2652/260, fol. 57ʳ.*

FIGURE 9 *Oronce Fine discussing the use of the astrolabe and astronomical tables with Urania. Detail of a woodcut in Oronce Fine,* De mundi sphaera *(Paris, 1542). Courtesy of the Adler Planetarium, Chicago.*

had one, "solidly and ingeniously worked in brass," which he bought with his own money.[50]

But as a precision tool, the astrolabe left something to be desired. Levi ben Gerson recognized some of the systematic and random errors involved in using an astrolabe to determine the altitude of a star: (a) the vertical diameter of the instrument might not be aligned with the zenith, thereby throwing off all altitude divisions around the limb; (b) the alidade's line of sight might not correspond with its reading on the scale; and (c) the user might read the scale to minutes even though its smallest subdivisions were degrees. The systematic errors (a and b) were introduced by faulty construction but could be rectified by new procedures for fabrication. The random error (c) was a consequence of method as much as design but could be lessened by the introduction of a transversal scale with finer subdivisions.[51] Few, however, took heed of Levi ben Gerson's recommendations. Although Tycho employed transversals on his mural quadrant and other large instruments, he did not incorporate them into any astrolabe, for he maintained that even the largest astrolabe could not measure the courses and positions of the stars with any sufficient degree of accuracy (or at least with the accuracy that Tycho demanded of his instruments). In comparison to other apparatus, an astrolabe was "neither…very convenient for observations of the stars, nor adequate and reliable." Indeed, the use of an astrolabe presupposed that the paths and positions of stars and planets were already determined by other methods.[52] That was why astronomers with astrolabes typically had ephemerides and other tables at their side.[53] (Figure 9)

If, as a handmaiden to theoretical astronomy, the astrolabe was not sufficiently fastidious, it was, nevertheless, uniquely suited for astrology. In noting this, Tycho took part in a long tradition. With the aid of planetary tables, an astrologer could use the instrument to determine

50. Tycho Brahe, *Astronomiae instauratae mechanica* (Wandesburgi, 1598), "De aliis quibusdam instrumentis nostris"; translated in Hans Ræder, Elis Strömgren, and Bengt Strömgren, eds. and trans., *Tycho Brahe's Description of His Instruments and Scientific Work* (Copenhagen, 1946), 99.

51. Goldstein (1977).

52. Ræder *et al.* (1946), 99–100.

53. See the image of an astronomer with an astrolabe and astronomical tables in a 13th-century manuscript (Paris, Bibliothèque de l'Arsenal, MS 1186, fol. 1v); and that in a manuscript composed between 1350 and 1399 (Copenhagen, Det Kongelige Bibliotek, Cod. Hebr. 37, fol. 114r), reproduced in Figure 3 (above).

the relationship of the planets to key stars and astrological houses.[54] On Latin instruments, the astrological houses were engraved directly on the tympans, while Persian astrolabes used unequal hour lines as the divisions between houses. That is not to say that Persian astrolabists had little interest in astrology, for astrological tables were abundantly inscribed on the backs of *mashriqi* instruments. As al-Bīrūnī observed in a book on the art of astrology (A.D. 1029/30), these included tables of lunar mansions, triplicities, and the relationship of the zodiacal signs to their faces, terms, and lords.[55]

The point of all this astrological furniture was to peep into the future. On one hand, astrologers practiced natural astrology, foretelling the weather, earthquakes, diseases, mortality, wars, discords, and conspiracies from the state of physical conditions, which were thought to be dependent both on the course of the planets through the signs and on the appearance of comets. On the other hand, they practiced judicial astrology, which attempted to divine the outcome of human affairs from the disposition of the heavens at the time of a client's birth or the occasion of an election. Judicial astrologers answered horary questions. In other words, they would be consulted by a person preparing to undertake a great project or go on a business trip. Once the astrologer learned the time of the client's birth, he saw how that person's horoscope corresponded with the aspect of the celestial bodies at the time the inquiry was made. Upon this comparison, he foretold how successful the undertaking would be — typically adding that God may do more or less as he pleased.

The principal appeal of the astrolabe in the late Middle Ages and the Renaissance was its handiness for astrology. In *The Canterbury Tales,* Chaucer gave the Oxford student, Nicholas, an astrolabe with which to practice his prognosticary art.[56] And 300 years later, John Aubrey remarked that every schoolboy should have a paper astrolabe "to teach him to erect a Scheme presently [*i.e.,* cast a horoscope]: wch will much delight & incourage them."[57] Astrolabes were used to determine a newborn's nativity, to find the propitious time to lay cornerstones, and to detect thieves (a common task of astrologers and cunning folk in villages).[58] Astrology was also a major component of medical care, and since the Middle Ages, astrolabes were used to determine the critical days of an illness, the optimum times for bleeding, and occasions for medications.[59]

Stars and human bodies came together in a memorable medieval report of astrolabes amidst the philosophers at Kublai Khān's table. These men were said to be wise in the sciences of astronomy, necromancy, geomancy, pyromancy, hydromancy, and augury:

> And every [one] of them have before them astrolabes of gold, some spheres, some the brainpan of a dead man, some vessels of gold full of gravel or sand, some vessels of gold full of coals burning, some vessels of gold full of water and of wine and of oil, some horloges of gold made full nobly and richly wrought, and many other manner of instruments after their sciences.[60]

54. On the use of the astrolabe to demarcate houses, find horoscopes, or set figures, see Stöffler (1513), part 2, props. 53-56; Köbel (1535), sigs. [C4]^r-D^r; Gemma Frisius (1556), chaps. 50-56; and Blagrave (1585), 42-47, 69-70. Henrion (1621), 62, recommends consulting ephemerides (alongside one's astrolabe) when wishing to determine astrological houses more exactly for the purpose of casting horoscopes.

55. Abū Rayḥān al-Bīrūnī, *Book of Instruction in the Elements of the Art of Astrology,* trans. Robert Ramsay Wright (London, 1934), 195; quoted in Turner, A. J. (1985), 22; Gibbs with Saliba (1984), 16, 32-38.

56. Chaucer (1987), *The Canterbury Tales,* "The Miller's Tale," A3209.

57. Oxford, Bodleian Library, MS Aubrey 10, fol. 109; quoted in Turner, A. J. (1973), 65.

58. Stöffler (1513), part 2, prop. 56; Stöffler (1560), part 2, prop. 56, fols. 175^r-183^r; Blagrave (1585), 47-48; Lynn Thorndike, *A History of Magic and Experimental Science,* 8 vols. (New York, 1923-58), 5: 259, 384. For the social functions of astrology and tasks such as thief detection, see Keith Thomas, *Religion and the Decline of Magic* (New York, 1971), chaps. 10-12.

59. Stöffler (1513), part 2, prop. 55; Franz Ritter, *Astrolabium* (Nuremberg, [1613]), see second part; also note the full title of Köbel (1535): *Astrolabii declaratio, eiusdemque usus mire jucundus, non modo astrologis, medicis, geographis, caeterisque literarum cultoribus multum utilis ac necessarius; verum etiam mechanicis quibusdam opificib. non parum commodus* (which describes the astrolabe as "very practical and even necessary for...doctors"). See also Lynn White, Jr., "Medical Astrologers and Late Medieval Technology," *Viator* 6 (1975): 295-308; reprinted in *idem, Medieval Religion and Technology: Collected Essays* (Berkeley, 1978), 297-315; and Nancy G. Siraisi, *Medieval and Early Renaissance Medicine: An Introduction to Knowledge and Practice* (Chicago, 1990).

60. Seymour (1968), 179. On the reliability of reports of the astrolabe in China, see discussion above, on pp. 4-6.

FIGURE 10 *Dr. Faustus conjuring in a chalk circle, with particular and universal astrolabes hanging on the wall. Woodcut from a chapbook in the collection of Samuel Pepys,* First Part of Dr. Faustus, Abreviated and brought into verse *(London, [*c. *1680]). Courtesy of the Master and Fellows, Magdalene College, Cambridge.*

While this passage, taken from *Mandeville's Travels,* may say more about medieval, Western expectations of the Orient than the real activities of the Great Khān's advisors, it should not be discounted as entirely fanciful. The passage may tell us something about the activities of diviners and philosophers in the service of Western potentates, or at least something about popular perceptions of sages among the sovereigns. The presence of astrolabists and astrologers in both Eastern and Western courts is not to be doubted.[61] Western sources, moreover, do confirm that astrolabes were used in the darker, prophetic arts. A medieval manuscript of Czech origin shows astronomers toiling with astrolabes alongside geomancers.[62] And in the *Liber particularis* (written at the turn of the thirteenth century), Michael Scot reported that knowledge of the astrolabe was won by Gerbert from demons whom he had conjured up and forced to teach him! Gerbert reputedly wrote down what he learned and shared his findings with others, later renouncing his contact with demons and becoming Pope.[63] Elsewhere in this text, Scot notes that the astrolabe itself was sometimes used in invoking evil spirits, but added that the Roman Church condemned this practice.[64] As late as the seventeenth century, astrolabes were popularly viewed as Faustian tools. Crude woodcuts in chapbooks (aimed at the lowest levels of the literate) depicted astrolabes alongside Dr. Faustus, seen conjuring in a circle, with the devil beside him.[65] (Figure 10)

61. See discussion above, on pp. 3, 6, and 7.

62. Krása (1983), plate 19.

63. Thorndike (1965), 93-94; *idem* (1923-58), 2: 322.

64. Thorndike (1965), 117. Scot believed that knowledge of astronomy was essential to necromancy, because there were proper times for convening with demons.

65. *First Part of Dr. Faustus, Abreviated and brought into verse* (London, [*c.* 1680]); Roger Thompson, ed., *Samuel Pepys' Penny Merriments* (New York, 1977), 100.

Astronomy and astrology were but two of the practical arts whose practitioners might find cause to consult an astrolabe. Geography, navigation, and surveying — which might be grouped together as topographical arts — were others. Medieval texts made frequent mention of geographical uses.[66] Any astrolabe could be used to find the difference in longitude between two towns if the initial time of a lunar eclipse was noted at each location.[67] Latitude was simply found by observing the altitude of the pole star or of the sun when it crossed the meridian (assuming one made adjustments for seasonal changes in solar declination).[68] Some astrolabes also included a geographical tympan with a map of the world in polar stereographic projection.[69] Features like these made the astrolabe a welcome companion on expeditions. John of Sacrobosco (fl. 1230-1255) pointed out, moreover, that if a person equipped with an astrolabe observed the pole star from two locations that were separated by 1° along a meridian, he could determine the circumference of the earth.[70] And according to *Mandeville's Travels*, itinerant astronomers were able to prove the earth to be round on the basis of such observations.[71]

The ability to find latitude with an astrolabe also made it increasingly attractive to navigators and explorers, who, in the Middle

66. Emmanuel Poulle in *Le navire et l'économie maritime du XVe au XVIIIe siècle* (Paris, 1967), 113; quoted in Francis Maddison, *Medieval Scientific Instruments and the Development of Navigational Instruments in the XVth and XVIth Centuries*, Agrupamento de Estudos de Cartografia Antiga, no. 30 (Coimbra: Junta de Investigações Científicas do Ultramar-Lisboa, 1969), 12 n. 35.

67. *Astrolabii...canones* (1512), sig. $b2^v$; Stöffler (1513), part 2, prop. 31; Jacquinot (1559), fols. 44^r-45^v; Danti (1569), 89-90.

68. *Astrolabii...canones* (1512), sig. $b2^r$; Stöffler (1513), part 2, prop. 30; Jacquinot (1559), fol. 43; Danti (1569), 87-88.

69. Gerard L'E. Turner and Elly Dekker, "An Astrolabe Attributed to Gerard Mercator, c. 1570," *Annals of Science* 50 (1993): 403-43, esp. 420. Also see Figure 5 (above) for an astrolabic map plate included in an astronomical compendium by Christopher Schissler.

70. The observations were used to mark the endpoints of an arc of 1°. The length of this arc was multiplied by 360 in order to compute the circumference. Lynn Thorndike, ed., *The Sphere of Sacrobosco and Its Commentators* (Chicago, 1949), 85, 122-23.

71. Seymour (1968), 139-40. Of course, altitude measurements of this sort only show the earth to be round from north to south. The times of lunar occultations or eclipses, observed simultaneously from different longitudes, can be used to show the earth to be round in the east-west direction.

FIGURE 11 *Mariner's astrolabe, dated 1616, recovered from the wreck of the* Atocha *off the Florida Keys. Courtesy of the Adler Planetarium, Chicago (A-275).*

FIGURE 12 *Mariner's astrolabe being used to measure the meridian altitude of the sun. Renaissance woodcut. Private collection.*

Ages, began to use it occasionally on shore.[72] Early use at sea would have been difficult, given the rocking of the ship and the wind resistance offered by the metal plate. These problems were overcome in the late fifteenth century, when the instrument was stripped of all nonessential parts, leaving only a heavily weighted, graduated limb and an alidade.[73] (Figures 11-12) In this new form, the astrolabe came of age as a nautical tool, and it was readily adopted by Renaissance sailors, who

72. See, for example, the likely use of an astrolabe made by a priest who returned to Norway in 1364 after a voyage to the "northern islands." R. A. Skelton, Thomas E. Marston, and George D. Painter, *The Vinland Map and the Tartar Relation* (New Haven, 1965), 180; cited in Maddison (1969), 12 n. 35.

73. Alan Stimson, *The Mariner's Astrolabe: A Survey of Known, Surviving Sea Astrolabes* (Utrecht, 1988). Also see a description of a Portuguese mariner's astrolabe (A-275, cat. no. 46), dated 1616, recovered from the wreck of the *Atocha*.

FIGURE 13 *Planispheric astrolabe being used for navigation. Painting of Noah's Ark,* c. *1590, attributed to Miskin. India, Mughal, Akbar period. Color and gold on paper. 28.1 x 15.6 cm. Courtesy of the Freer Gallery of Art, Smithsonian Institution, Washington, D.C. (48.8).*

were taught to sail by the altitude of the pole star or "run down the latitude."[74]

There is still some question whether the traditional astrolabe ever went to sea.[75] A chart drawn by Diego Ribero in 1525 arguably contains the earliest illustration of a seaman's astrolabe, and it resembles not the mariner's astrolabe in its typical, open form, but the solid back of the traditional planispheric astrolabe.[76] A traditional astrolabe also appears on board ship in an Indian painting from the late sixteenth century.[77] (Figure 13) Texts and instruments further muddle the issue. In a treatise on the planispheric astrolabe first published in 1513 and reprinted many times, Johann Stöffler told explorers how to use the standard instrument to find their location if they went astray in the middle of the sea or in the wilderness; and in another text, John Blagrave taught how to pilot a ship by the stars.[78] Gemma Frisius, moreover, designed a *quadratum nauticum* for mariners, which his nephew, Gualterus Arsenius, included within the mater of many planispheric astrolabes.[79] The *quadratum nauticum* indicated the rhumb — *i.e.*, the course given by a magnetic compass — to be sailed between places of known latitude and longitude.[80] We must ask why Arsenius went to the trouble to put this device on his astrolabes. Perhaps it was sheer fancy for him to think that his magnificent instruments would ever go to sea, or perhaps he thought that the *quadratum* would call attention to the recent evolution of the astrolabe into a nautical instrument. Another possibility may be that the enhanced astrolabe appealed to the armchair traveler!

Whatever the case, the fact that Arsenius also included astrological notations on the star-pointers of his retes reminds us that astrolabists stood at the nexus of astrology, astronomy, and nautical science. No story illustrates this better than that of Abraham Zacuto, the maker of scientific instruments, compiler of astronomical tables, and professor of astronomy and astrology at the University of Salamanca. In 1496, prior to the departure of Vasco da Gama, the king of Portugal asked

74. In sailing by the altitude of the pole star, seamen used the height of Polaris above the horizon as a measure of their vessel's change in the north-south position. Once Europeans ventured south of the equator (in 1471), it became impractical to sight Polaris. For mariners near the equator, Polaris appeared too close to the horizon; for those in southern waters, Polaris was below the horizon. The technique of altitude sailing was then refined into the method of "running down the latitude." According to this method, sailors measured the altitude of either the sun or pole star in order to determine their latitude. Based on this, they sailed north or south until they reached the latitude of their destination and then ran east or west along that parallel until they sighted land. On use of the mariner's astrolabe by Vasco da Gama, see Maddison (1969), 28; and the marginalia of Christopher Columbus in his copy of Pierre d'Ailly, *Ymago mundi* (Louvain, [*c.* 1480-83]); trans. and printed by Grant (1974), 636 n. 42. On navigation, see David W. Waters, *Science and the Techniques of Navigation in the Renaissance*, 2nd ed. (Greenwich, 1980); E. G. R. Taylor, *The Haven-Finding Art* (New York, 1957); and David W. Waters, *The Art of Navigation in England in Elizabethan and Early Stuart Times*, 2nd ed. (Greenwich, 1978).

75. Emmanuel Poulle, for instance, thinks it did not; see Maddison (1969), 12-13.

76. Maddison (1969), 28-29.

77. Painting of Noah's Ark, attributed to Miskin (India, *c.* 1590), accession no. 48.8, Freer Gallery of Art, Smithsonian Institution, Washington, D.C.

78. Stöffler (1513), part 2, prop. 33; Blagrave (1585), 62.

79. Gemma Frisius (1583), chap. 87; Apianus (1584), 55, 459-62. See astrolabes by Arsenius, dated 1558 (M-23, cat. no. 8) and 1564 (M-24, cat. no. 9), and the unsigned Louvain astrolabe, *c.* 1600 (M-25, cat. no. 10), in this volume.

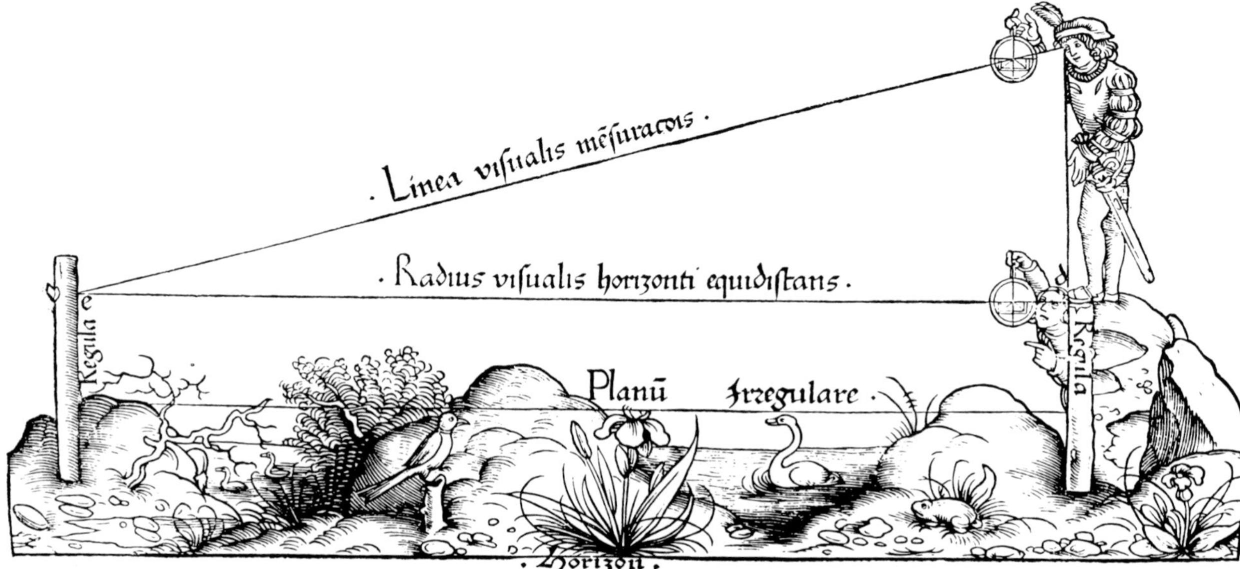

FIGURE 14 *Use of an astrolabe to find the distance between inaccessible places. From Johann Stöffler,* Elucidatio fabricae ususque astrolabii *(Oppenheim, 1524). Courtesy of the Adler Planetarium, Chicago.*

80. The *quadratum nauticum* was a windrose within a latitude-longitude grid. It was useful to navigators who wanted to sail between places with known geographic coordinates. The navigator marked the difference in longitude along the horizontal scale and the difference in latitude along the vertical. He then plotted the intersection of these lines within the windrose in order to discover the rhumb to follow. This technique was good only for short distances, as Gemma himself remarked. Michel (1947), 44-45.

81. Metzger and Metzger (1982), 157; *Encyclopaedia Judaica, s.v.* "Zacuto, Abraham ben Samuel."

82. Abū Rayḥān al-Bīrūnī, *The Exhaustive Treatise on Shadows*, trans. E. S. Kennedy, 2 vols. (Aleppo, 1976); quoted in Turner, A. J. (1985), 21. The shadow square is sometimes attributed to al-Battānī (850-929); see Edmond R. Kiely, *Surveying Instruments: Their History* (Columbus, Ohio, 1979), 68.

83. Kiely (1979), 74; Grant (1974), 180-82.

84. *Astrolabii...canones* (1512), sigs. [b8r]-d3r; Stöffler (1513), part 2, props. 58-65; Copp (1525), sigs. fijr-giijv; Köbel (1535), sigs. D2r-[G]r; de Rojas (1550), 165-81, 185-96. For an example of an all-purpose, applied-mathematics book tutoring surveyors, t pographers, architects, and artists on the use of the astrolabe, see Cosimo Bartoli, *Del modo di misurare le distantie, le superficie, i corpi, le piante, le provincie, le prospettive, & tutte le altre cose terrene, che possono occorrere a gli huomini* (Venice, 1564). This astrolabe was so identified with the work of surveyors and architects that one appears in Sir John Harington, *A New Discourse of a Stale Subject, Called the Metamorphosis of Ajax* (London, 1596). In this mock-heroic work, Harington described his invention of a water closet. On one page, we see a "rare Engine[e]r" named "Archimides" using an astrolabe to survey the placement of "a jakes," or privy, of the new form.

Zacuto not only to train sailors in the use of his astrolabe, astronomical tables, and charts, but also to predict the outcome of the expedition.[81]

While finding one's latitude was an important step in navigation, it was also critical for laying a base line in a geodetic survey. The astrolabe was seldom (if ever) used for this task — its limb being too crudely divided — but it was well-adapted for local surveys of territories filled with landmarks. For this purpose, the surveyor took advantage of a device known as the shadow square found on the back of the mater. The shadow square, which has been attributed to al-Khwārizmī (d. post-A.D. 847/8), was to be found very early on Islamic instruments.[82] In the West, Gerbert was the first to describe its practical use, and medieval treatises on applied geometry, such as that written by Dominicus de Clavasio (fl. 1346), commonly included problems involving the use of the shadow square to determine the distance between places, the height of buildings, or the depth of wells.[83] The same problems were discussed and whimsically illustrated in Renaissance texts that promoted the astrolabe as a tool for the surveyor, architect, and artist.[84] (Figures 14-17)

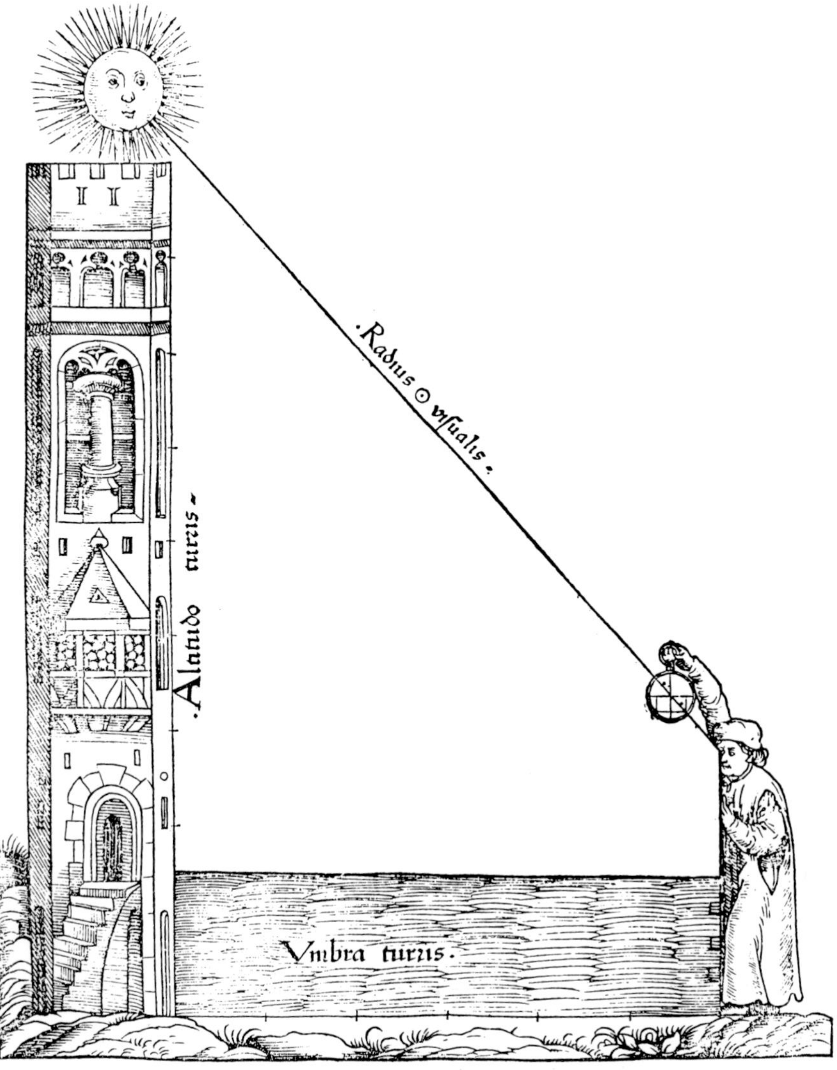

FIGURE 15 *Use of an astrolabe to find the height of a tower. From Johann Stöffler,* Elucidatio fabricae ususque astrolabii *(Oppenheim, 1513). Courtesy of the Adler Planetarium, Chicago.*

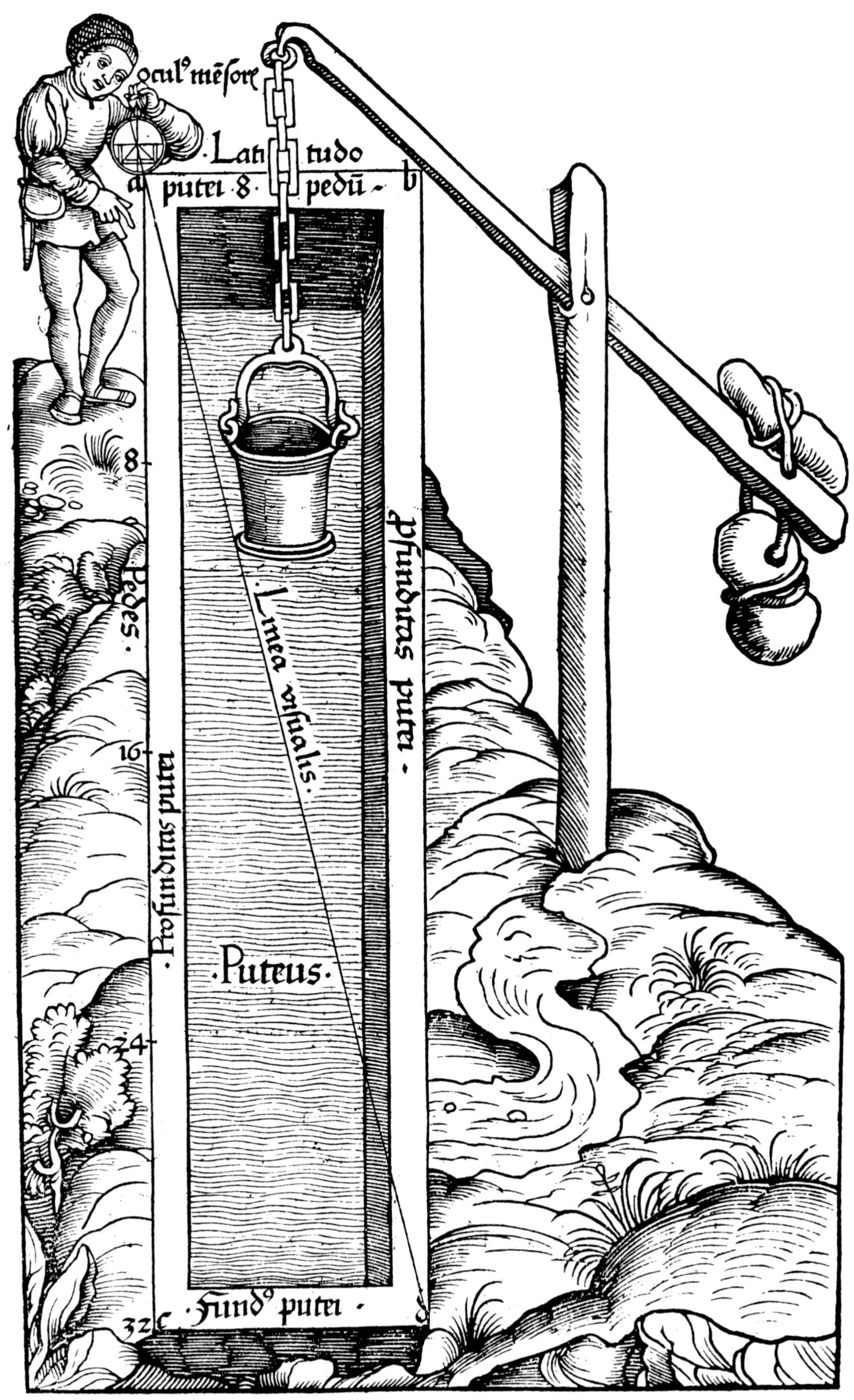

FIGURE 16 *Use of an astrolabe to find the depth of a well. From Johann Stöffler,* Elucidatio fabricae ususque astrolabii *(Oppenheim, 1513). Courtesy of the Adler Planetarium, Chicago.*

Another innovation relevant to surveying occurred in 1486, when Hans Dorn inserted a magnetic compass in the throne of an astrolabe he made for Martin Bylica.[85] This enabled angles of azimuth to be measured along the limb of the astrolabe when the instrument was laid on its side. Gemma Frisius independently endorsed this idea and championed the astrolabe as a tool for land measure and triangulation.[86] (Figure 18) Even without the compass, the astrolabe was helpful in rendering maps and topographical reports.[87] (Figures 19-20) It was further promoted as an agent of geometricized warfare, for it could be suspended from a spear and used to determine whether a distant enemy was advancing or retreating.[88] (Figure 21) The surveyor and military tactician, however, had little need for the stellar rete, tympans, and astrological lines on an astrolabe, and their instruments became stripped of these luxury features.[89] In this way, more specialized apparatus for surveying and navigation evolved from the astrolabe during the Renaissance, and these new tools ultimately superseded it in those fields.

85. The astrolabe is pictured in Turner, A. J. (1987), 36-37; and Turner, A. J. (1985), 48 n. 164. The Italian astrolabe, *c.* 1650 (M-30, cat. no. 24), has a compass in its throne, and the de Rojas-type astrolabe, German?, 17th century (M-42, cat. no. 17), likely did as well.

86. Gemma Frisius's *Libellus de locorum describendorum ratione* (Antwerp, 1533) is printed with an accompanying discussion in de Rojas (1550), 201-25; see also Gemma Frisius (1583), 97-98; and Alexander Pogo, "Gemma Frisius, His Method of Determining Differences of Longitude by Transporting Timepieces (1530), and His Treatise on Triangulation (1533). With...a Facsimile Reproduction (No. XVI) of Gemma's *Libellus de locorum describendorum ratione*, Antwerp, 1533," *Isis* 22 (1934): 469-504.

87. De Rojas (1550), 182-85, 197-98; Bartoli (1564), fols. 43^{V}-46^{V}; Henrion (1620), 2.

88. De Rojas (1550), 199-200; Bartoli (1564), fol. 49.

89. Surveyor's astrolabes will be treated in the surveying section of the comprehensive catalogue of the Adler collection. Examples from the third quarter of the 16th century include M-43, M-44, and M-45 (cat. no. 13).

FIGURE 17 *Use of an astrolabe to discover the proportions of a building. From Cosimo Bartoli,* Del modo di misurare *(Venice, 1614). Courtesy of the Adler Planetarium, Chicago.*

As the astrolabe slipped from favor in Christian Europe, it was still prized in Islamic communities because it played a larger social role. It was not simply a tool for astronomy and related land-based sciences (such as navigation and surveying), but an instrument used to bolster faith and ritual.

This had also been true to a modest extant in Judeo-Christian circles. In Biblical exegesis during the twelfth and thirteenth centuries, Jewish scholars argued whether cryptic words, which seemed to indicate

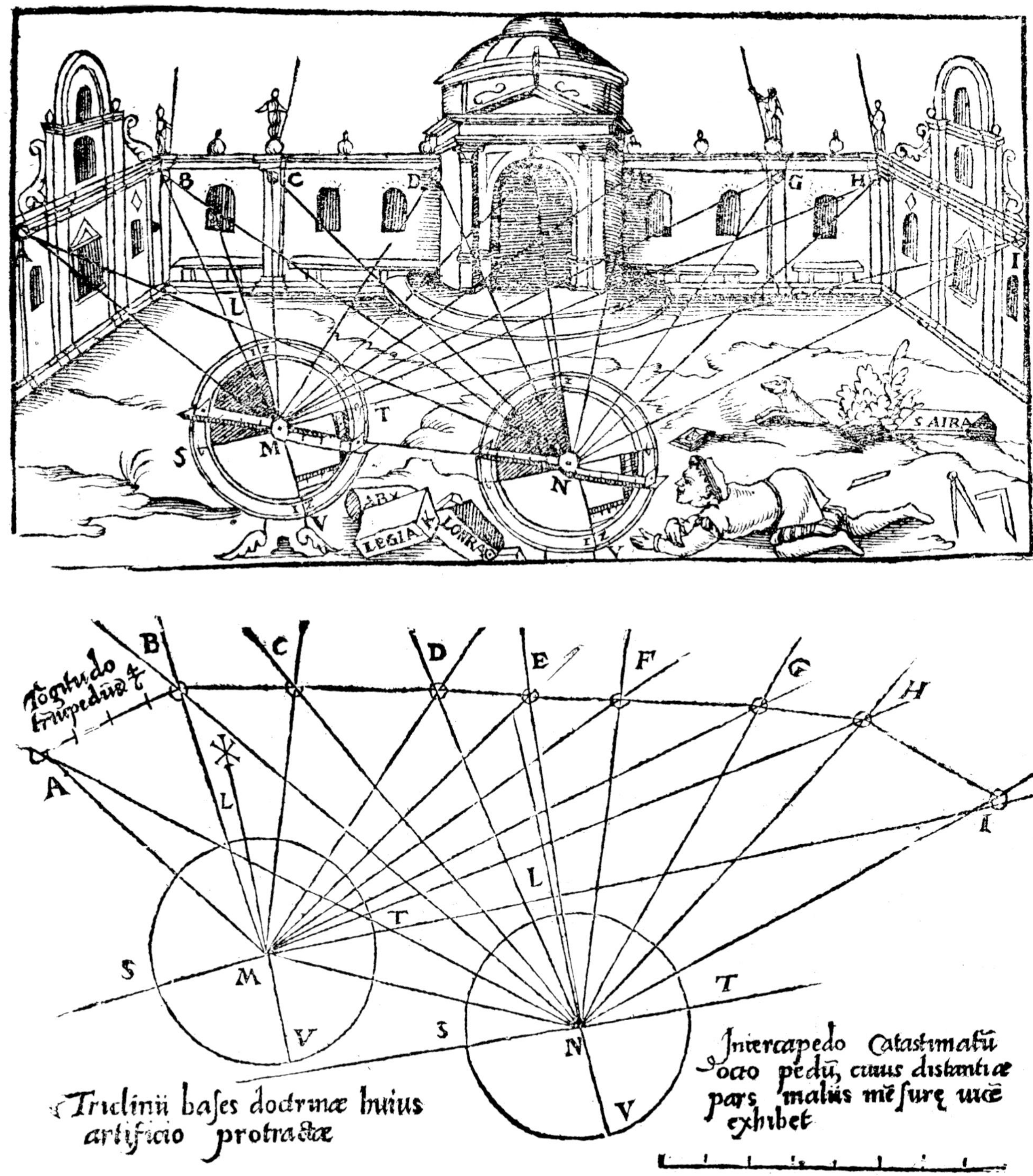

FIGURE 18 *Triangulation with an astrolabe positioned at two stations. From Gemma Frisius,* De astrolabo catholico liber *(Antwerp, 1583). By permission of the Houghton Library, Harvard University.*

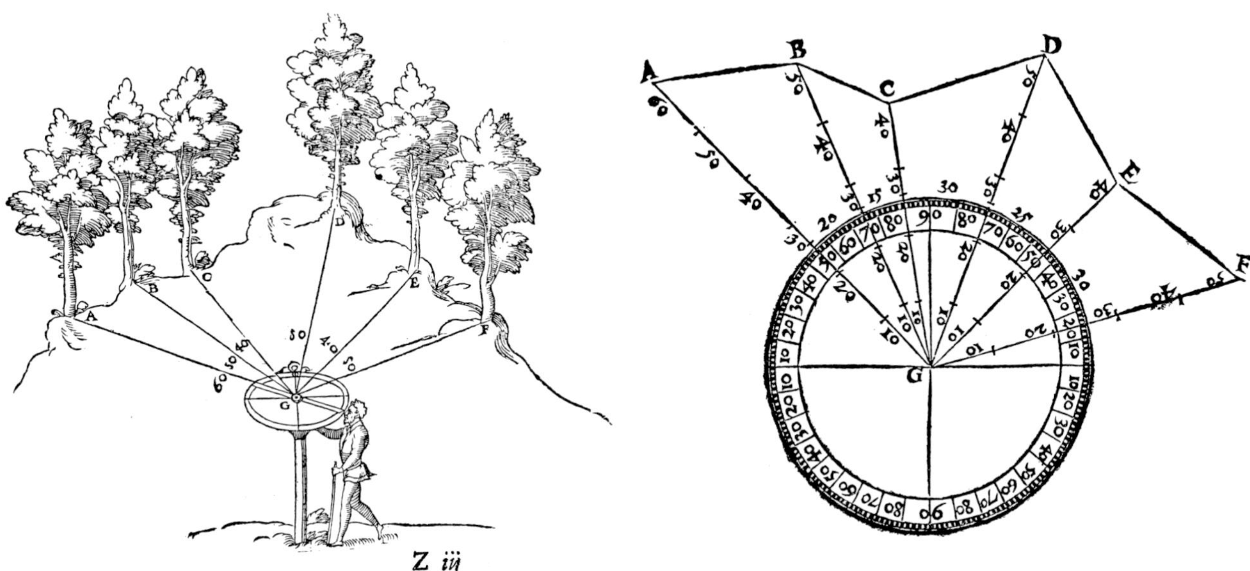

FIGURE 19 *Use of an astrolabe to map the position of some trees: A surveyor observes the angles between the trees and measures the distance between each tree and a reference point. Then, with a protractor and ruler, he reproduces his field measurements to scale on paper, and so constructs a map. From Juan de Rojas,* Commentariorum in astrolabium quod planisphærium vocant, libri sex *(Paris, 1551). Courtesy of the Adler Planetarium, Chicago.*

divinatory devices, might allude to an astrolabe.[90] For instance, Abraham ibn Ezra (*c.* 1090-*c.* 1164), a scholar renowned for his treatises on the astrolabe and works on astronomy as much as for his commentary on the Bible, apparently suggested that the mysterious *urim vethumin* — held in the breastplate of the High Priest and revealing divine will when consulted — referred to an astrolabe.[91] Rabbis also debated whether it was permitted to use an astrolabe on the Sabbath. Solomon ibn Adreth allowed it because it was equal to reading a scientific book; others permitted it because they believed that Rabban Gamliel II (fl. *c.* A.D. 80-116) had used one on the Sabbath.[92]

Christians sometimes made their astrolabes into tools of religion, too. They inscribed tables of Saints' Days, dominical letters, epacts, and the paschal index on the back of the mater so that they could readily determine the date of Easter and other religious festivals.[93] Nevertheless, such tables were added more as an afterthought; they were nice to have, but were not the main reason one owned an astrolabe.

By contrast, religion was a driving force in the acquisition and use of astrolabes in Islamic communities, where every *muwaqqit* (the astronomer at the local mosque) might use one in determining the times of the five daily prayers — namely, sunset, nightfall, daybreak, midday, and mid-afternoon.[94] In fact, mosque astronomers were sometimes drawn from the ranks of astrolabists, and various devices adapted the astrolabe to their needs.[95] Arcs for determining the hours of prayer from the altitude of the sun were inscribed on the back of *mashriqi* instruments, whereas prayer lines were placed on the tympans of *maghribi* astrolabes. On instruments from both eastern and western Islam, circular cotangent scales could also be used to determine the times of afternoon prayer from the ratio of a gnomon's length to the length of its shadow. Twilight lines occasionally appeared on the tympans, too, dawn and nightfall being times for prayer.[96]

90. Texts of particular interest included Genesis 31:19, Exodus 28:30, and Numbers 22:7; see *The Soncino Chumash: The Five Books of Moses with Haphtaroth*, Hebrew text with English translation and commentary, ed. A. Cohen (London, 1947); and Solomon Gandz, "The Astrolabe in Jewish Literature," *Hebrew Union College Annual* 4 (1927): 469-86, esp. 480-81.

91. Ibn Ezra's commentary on Exodus 28:30; discussed in Gandz (1927), 471-72, 480. On ibn Ezra's scientific work, also see Millàs-Vallicrosa (1963), 151-52.

92. Gandz (1927), 481-82. For discussion of Hebrew astrolabes, see Bernard Goldstein, "The Hebrew Astrolabe in the Adler Planetarium," *Journal of Near Eastern Studies* 35 (1976): 251-60.

93. See the English astrolabe, *c.* 1250 (M-26, cat. no. 1), the Martinot astrolabe, 1598 (M-31, cat. no. 15), and the Danfrie-Moreau astrolabe, 1584-1622 (W-98, cat. no. 19) in this volume.

94. King (1987a), 2: 105. I am speaking here of people working in the tradition of Islamic mathematical astronomy. Folk astronomy was also concerned with the religious aspects of Muslim daily life, but writers on folk astronomy rarely, if ever, recommended astrolabes, quadrants, sundials, or tables, which were the standard instruments of the mosque astronomer. Instead, they advocated crude formulas and simple procedures for regulating the times of prayer or finding the direction of Mecca. For a comparison of mathematical and folk astronomy, see David A. King, *Astronomy for Landlubbers and Navigators: The Case of the Islamic Middle Ages*, Centro de Estudos de História e Cartografia Antiga, no. 164 (Lisbon: Instituto de Investigação Científica Tropical, 1984).

95. Mayer (1956), 40-41, 61.

96. Gibbs with Saliba (1984), 31-33, 38-39, 51, 54; Hartner (1968/1984), 1: 299.

FIGURE 20 *Use of an astrolabe to map the placement of some statues. From Cosimo Bartoli,* Del modo di misurare *(Venice, 1614). Courtesy of the Adler Planetarium, Chicago.*

Devout Muslims, moreover, were not only required to pray at certain times of day, but were expected to face Mecca. Hence, *mashriqi* astrolabes typically had arcs representing the azimuth of the *qibla* for key cities, and these were employed in finding the direction of Mecca from the altitude of the afternoon sun.[97] Persian and Indo-Persian astrolabes also contained gazetteers, which listed cities, their latitudes, longitudes, and other geographic parameters pertaining to their orientation to Mecca. Such parameters included the *inḥirāf* (the azimuth of the *qibla*) or *jihat* (the direction of the azimuth of the *qibla* with respect to the four cardinal points). As might be guessed, Mecca and Medina were often given pride of place in the gazetteer.[98]

By the seventeenth century, the throne of Persian instruments was sometimes inscribed with verses from the Koran containing the word *"kursī,"* which is Arabic for "throne."[99] This was tantamount to a pious dedication, and appropriate for an instrument so imbued with religious purpose.

Prized for its utility, convenience, and beauty, the astrolabe was the most important astronomical instrument of the Middle Ages and early Renaissance, and is justly famous today. It is appropriate that the first two volumes of the catalogue of scientific instruments in the History of Astronomy Collection at the Adler Planetarium be devoted to this sophisticated instrument.

97. *Qibla* is the Arabic term denoting the direction of Mecca, or to be precise, the direction of the *Ka'ba,* the most important sanctuary of Islam, situated near the center of the great mosque in Mecca. Muslims around the world direct their prayers to this sanctuary. See David A. King, "Ḳibla," in *Encyclopedia of Islam*, new ed., 5: 83-88 (1979b).

98. Gibbs with Saliba (1984), 26-33; Hartner (1968/1984), 1: 303.

99. Gibbs with Saliba (1984), 22.

100. Welborn (1931), 191.

Let me conclude with a short story: In 1025, Radulf of Liège wrote to a school friend, Ragimbold of Cologne, inviting him to visit during the festival of St. Lambert, in order to see his astrolabe. "You will not be sorry for it," he added.[100] As you begin to peruse this catalogue of astrolabes, the curatorial staff of the Adler Planetarium joins me in saying, like Radulf, that it will be well worth your while.

FIGURE 21 *Use of an astrolabe to survey enemy lines. From Cosimo Bartoli,* Del modo di misurare *(Venice, 1614). Courtesy of the Adler Planetarium, Chicago.*

Astrolabes

FIGURE 1 *The Armillary Sphere. Sacrobosco (1485), frontispiece.*

The Astrolabe: A Technical Introduction

by Marjorie and Roderick Webster

The astrolabe is one of the earliest and, at the same time, one of the most sophisticated of all ancient scientific instruments. It is primarily a computer. With it the time of day or night can be found by observing the altitude of the sun or a star above the horizon. A traveler may find his direction at all times of the year, as the azimuth of the sun or a star is also computed from its altitude. In addition, the astrolabe graphically shows the positions of the sun, the stars, and, with the aid of an almanac, the moon and the planets at any moment of the year, so that their rising and setting times can be determined.

The astrolabe derives its name from the Greek word *astro*, meaning star, plus the word *labio*, meaning taker, finder, thief. Thus, literally, it is a star finder.

The classic astrolabe, by far the most common form of this instrument, was drawn using a south stereographic projection to map the stars and the other coordinates. The South Celestial Pole was chosen as the center of the projection, and the plane that runs through the celestial equator as the plane of reference. (Figure 2) Thus a man

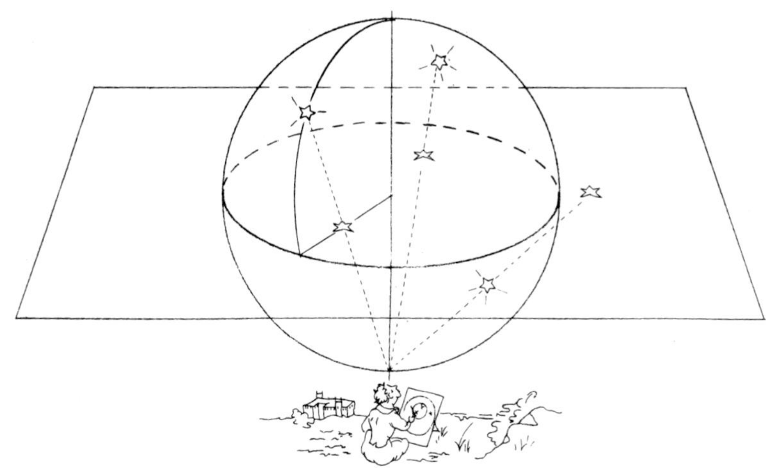

FIGURE 2 *The Stereographic Projection. Webster (1984), 6.*

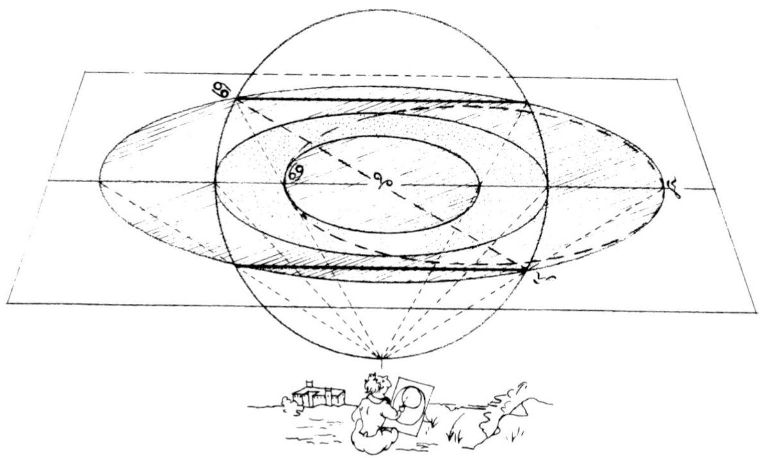

FIGURE 3 *The Projection of the Tropics. Webster (1984), 7.*

with his eye at the South Celestial Pole could, with a long enough pointer, mark on this plane the apparent location of any star.

The stereographic projection was a brilliant choice: the angle between any three stars in the sphere will be the same as the angle between the projection of these stars onto the plane, and any circle on the celestial sphere will be represented on the plane as a circle. In this way it is possible to produce a flat map of the heavens in which the angular relationship between the stars is maintained.

In Figure 3 it will be seen that the projection of the Tropic of Capricorn forms the largest circle drawn on the plane of the equator; the Tropic of Cancer forms the smallest. Connecting the circles of Capricorn and Cancer, and shown as a dotted line, is the projection of the ecliptic circle. In this diagram we see the basic form of the classic astrolabe.

This type of astrolabe is made up of two projections of the celestial sphere. The first is the rete, which is an openwork star map. (Figure 4) The positions of the brightest stars are indicated by named pointers, and the sun, the moon, and the planets can be located on the ecliptic circle, which is also shown. The center pivot of the rete marks the position of the North Celestial Pole. The second projection is the tympan. (Figure 5) This plate, drawn for a specific latitude, shows the principal coordinates in relation to the observer, *i.e.,* the horizon, the lines of equal altitude (almucantars), the zenith, the meridian, and the azimuths, or direction lines. It is locked into the the body of the instrument, known as the mater, with a tang to keep it from rotating.

It is customary to have a number of tympans in an astrolabe, each one figured for a different latitude. These and the rete are contained in the mater, which means "mother" or "womb." The rim or edge of the mater, also known as the limb, is usually divided into 24 equal hours, numbered twelve to twelve twice. This hour scale completes the information necessary to solve the problems of time from the positions of the stars.

The mater is equipped with a ring and shackle arranged so that the instrument will hang vertically when it is being used for measuring the altitude of the sun or a star. (Figure 6) All observations are made using the back of the astrolabe, where the outer edge is divided into degrees and a sighting bar, or alidade, is pivoted at the center.

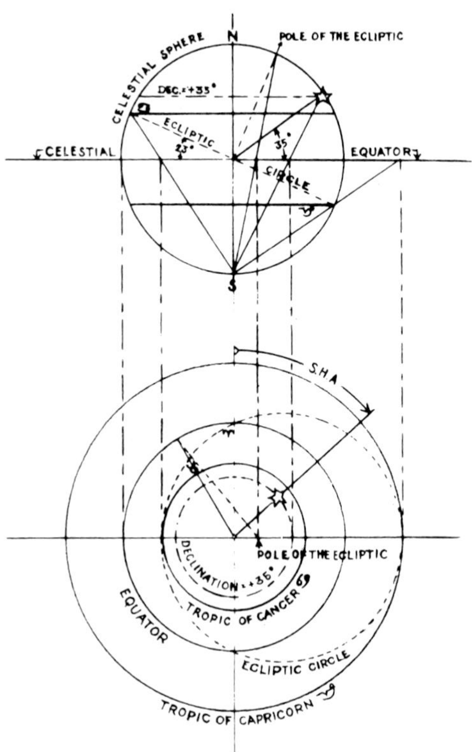

FIGURE 4 *Geometric Construction of the Rete. Webster (1984), 8.*

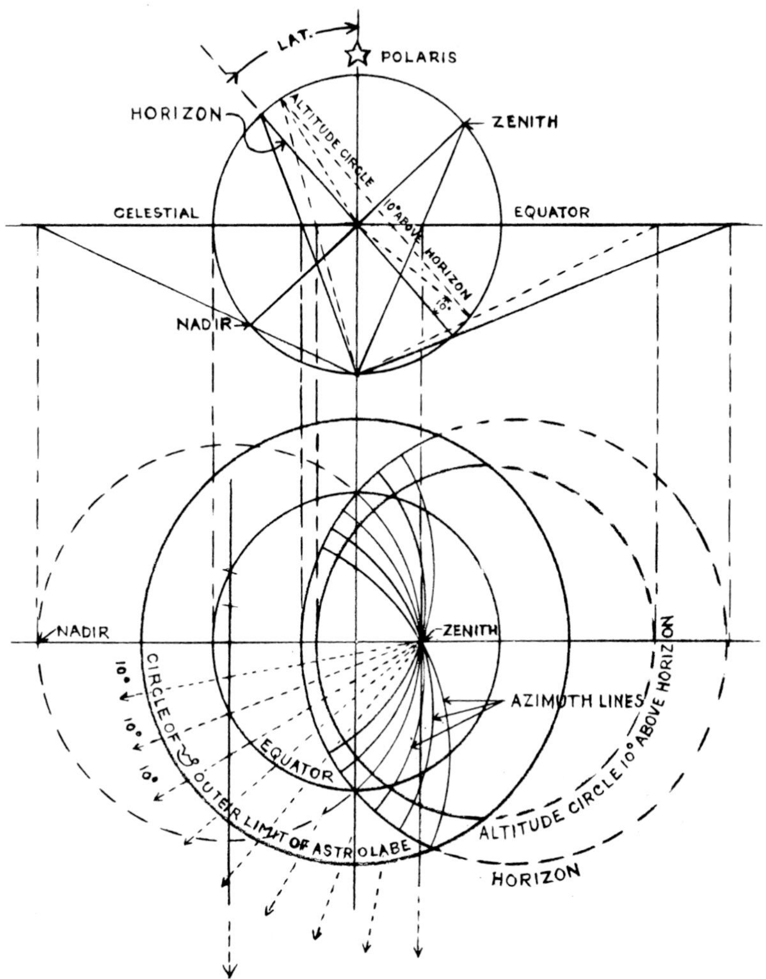

FIGURE 5 *Construction of the Tympan. Webster (1984), 9.*

Also on the backplate of most classic astrolabes are two circular calendars, one showing the zodiac, the other the civil calendar. The zodiac is placed outside, next to the limb. It is divided into 360 degrees and is usually marked 0°-90°-0° twice, with the zero points on the horizontal center line of the astrolabe and the 90° marks on the vertical meridian. These marks are used in connection with the alidade, to measure the altitude of the sun or a star.

The same circle is also divided into twelve parts of 30 degrees each, which represent the signs of the zodiac; they are labeled with the name and/or symbol of each. The degree divisions are marked 10-20-30 within each of the twelve segments.

The civil calendar, which is drawn inside the zodiac, is divided into months and days, each month showing 28, 30, or 31 days as appropriate. The two circular scales are so arranged that the position of the sun in the zodiac may be found opposite the date selected from the civil calendar. This concordance will only be accurate for the year in which the astrolabe was made, but the change is slow and it takes many years to build a significant error.

The two circular calendars may be either concentric or eccentric. (Figures 7 and 8) If they are concentric, the 365 day divisions will be slightly irregular due to the elliptical orbit of the earth. If they are eccentric, the day circle may be divided equally.

FIGURE 6 *Man Sighting the Sun. Jacquinot (1625), 265.*

The unequal, or planetary, hours are shown on the tympan. In this method of reckoning time, which was usual in Europe until the sixteenth century and in Japan until the nineteenth century, the periods of daylight and darkness are always divided into twelve hours each, so that day and night hours are unequal except at the equinoxes. To find the unequal hour, the rete is set from an observation of the sun or a star, and the rule, which lies over the rete, is set to the position of the sun in the zodiac. The far end of the rule will cross the ecliptic circle at a point 180 degrees from the sun's position. This point will lie in one of the unequal hours, which are shown below the horizon on the tympan, and will indicate the planetary or unequal hour. (Figure 10)

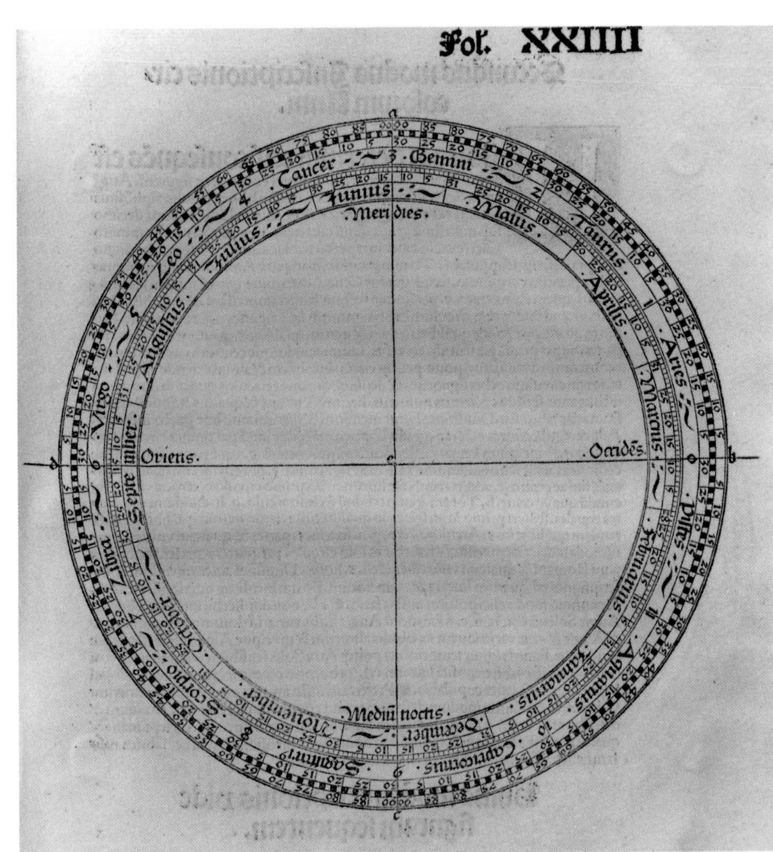

FIGURE 7 *Concentric Calendar Scales. Stöffler (1513), fol. xxiiii*[r]*.*

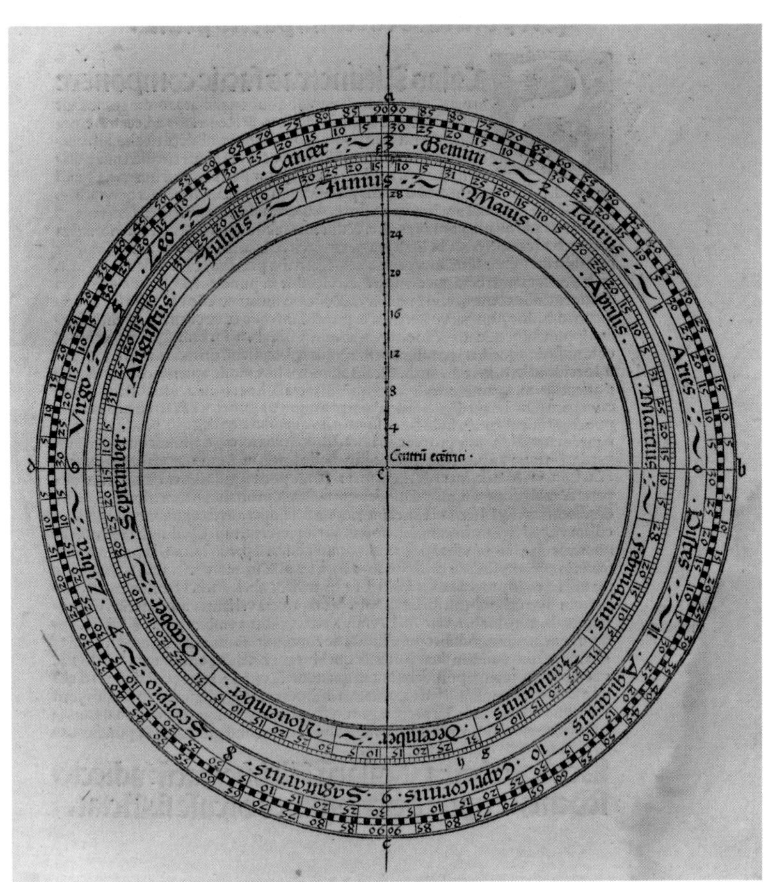

FIGURE 8 *Eccentric Calendar Scales. Stöffler (1513), fol. xxv*[r]*.*

Parts of the Classic Astrolabe

THE MATER

The mater, or main body, of the astrolabe is hollowed out to hold the rete and several tympans, which are drawn for different latitudes. A scale of hours is engraved on the rim, and a shackle and ring are provided to hold the mater. (Figure 9)

THE TYMPAN

The tympan, or plate, is engraved with the coordinates of the sphere. These include the zenith, the horizon, and the lines of altitude (almucantars) and azimuth, all of which are drawn for a specific latitude. The Tropics of Cancer and Capricorn are also shown. These last are the same for all latitudes. Tympans usually carry a diagram of unequal hours, which is used to convert the equal hours found with the astrolabe to the older unequal hours. Many also show the Great Houses used by astrologers. (Figures 10 and 11)

THE RULE

The rule, which lies over the rete, is used to line up the date on the ecliptic circle with the correct time on the hour circle. It often carries a scale showing declination to aid in reading the unequal hour diagram. (Figure 12)

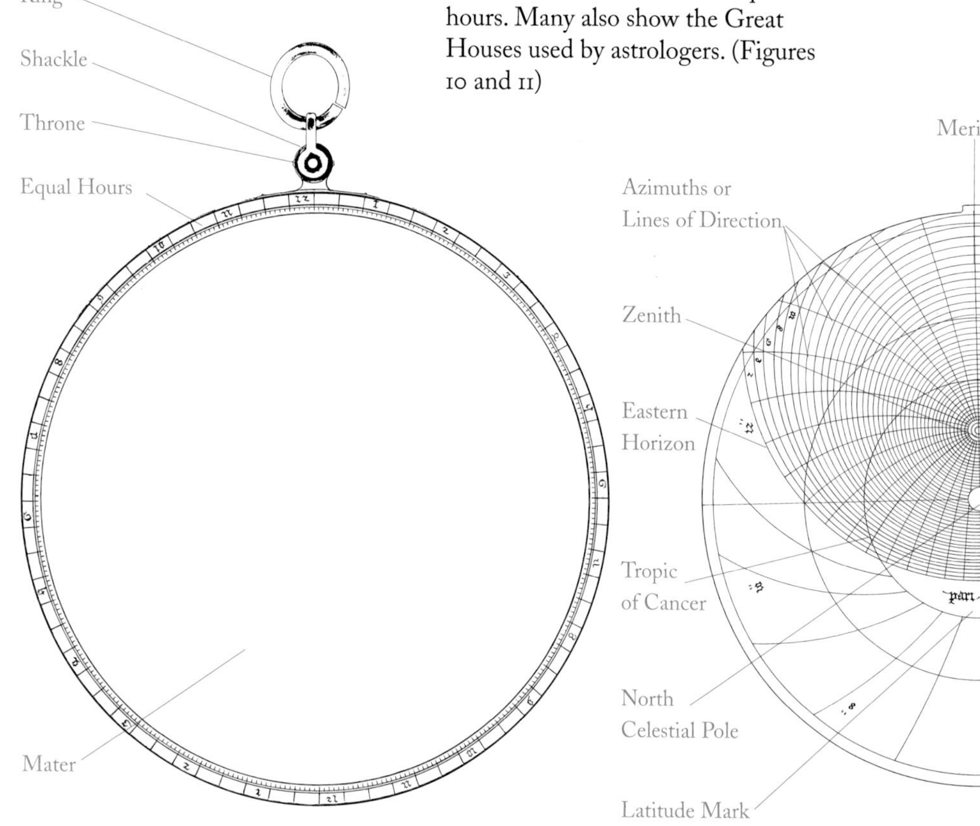

FIGURE 9 *The Mater. Webster (1984), 10.*

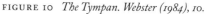

FIGURE 10 *The Tympan. Webster (1984), 10.*

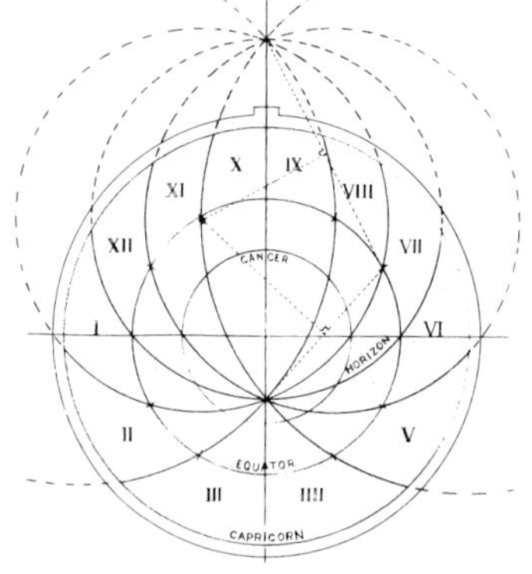

FIGURE 11 *The Great Houses. Webster (1984), 14.*

THE RETE

The rete is the star map. The center pivot marks the position of the North Celestial Pole. Many bright stars are indicated by named pointers, and the paths of the sun, moon, and planets are shown by the ecliptic circle. This circle is divided into the twelve signs of the zodiac on which the date settings are made. (Figure 13)

THE ALIDADE

The alidade, or sighting bar, is centrally pivoted on the back of the astrolabe. It is equipped with pinhole sights and is used with the scales engraved on the backplate. The whole instrument is held together by a pin and a wedge; the latter, which traditionally was shaped like a reclining horse, is therefore referred to by that name. (Figures 14A and 14B)

THE BACK OF THE MATER

All observations and measurements are made using the back of the astrolabe. A circle of degrees engraved around the edge is used to measure the altitude of the sun or a star. Another scale, the shadow square, is especially useful to the surveyor. The calendar scales usually complete the furniture on the back of the mater. (Figure 15)

FIGURE 12 *The Rule. Webster (1984), 11.*

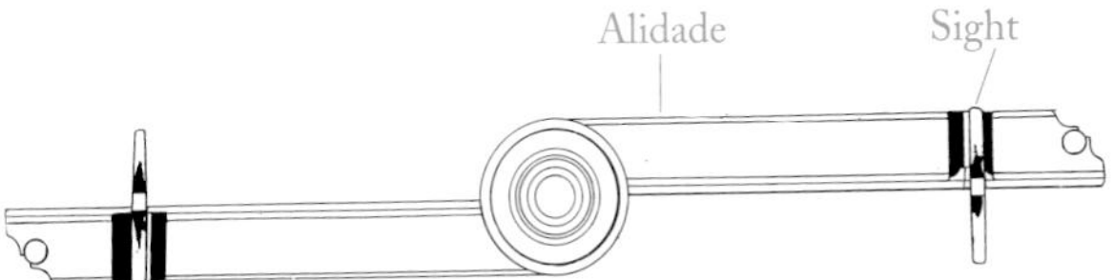

FIGURE 14A *The Alidade. Webster (1984), 11.*

FIGURE 14B *Pin and Horse. Webster (1984), 11.*

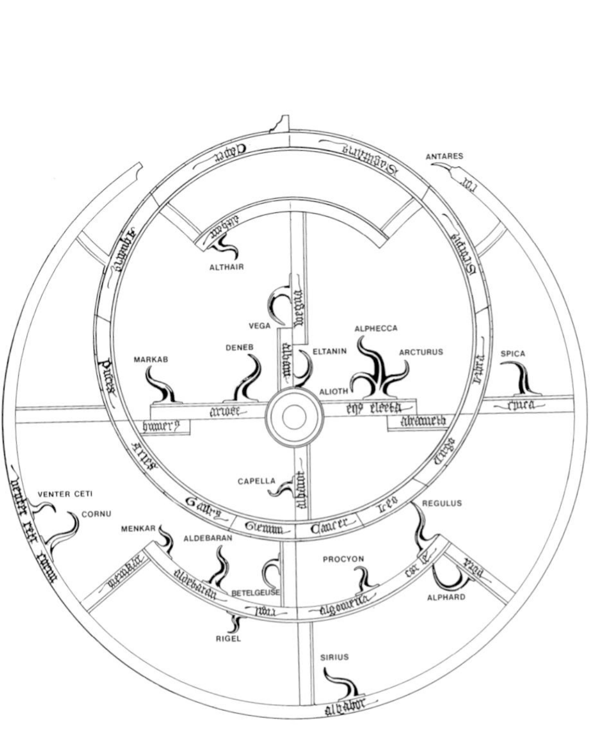

FIGURE 13 *The Rete. Webster (1984), 11.*

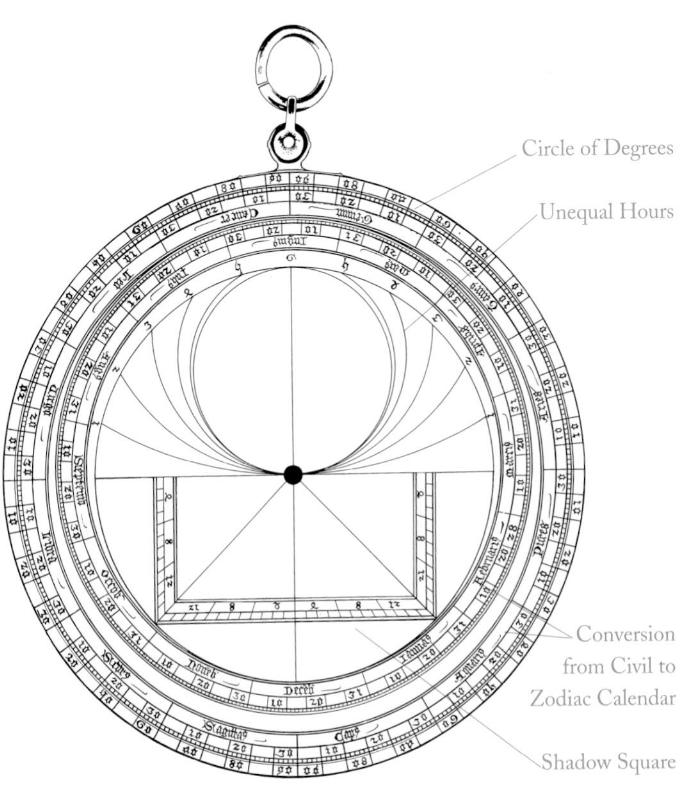

FIGURE 15 *The Back of the Mater. Webster (1984), 3.*

Other Types of Astrolabes

UNIVERSAL ASTROLABES

To avoid the need for many tympans, each engraved for a specific latitude, a number of astrolabes were developed that could be used in any location; thus they are known as universal astrolabes. One of these, the spherical astrolabe, is represented by a unique example in the Museum of the History of Science at Oxford.[1] (Figure 16)

There are also planispheric universal astrolabes, which achieved wide acceptance. They all show a figure of a transparent celestial sphere viewed from a point on the line determined by the vernal and autumnal equinoxes and marked with meridians and parallels. The point of projection, however, varies from one designer to another. All show both the North and South Celestial Poles. The first point of Aries and the first point of Libra appear in the center of the projection. Since both the front and the back of the sphere are visible on the same plane, care must be taken to find which hemisphere a star is located in. The meridians are used to plot the Right Ascensions of the stars and the sun, and the difference between them is used to determine the time.

THE SPHERICAL ASTROLABE This astrolabe was described in the *Libros del saber* of Alfonso X of Castile (1221-1284).[2] The stars and ecliptic circle are shown on a rete that takes the form of a cut-out spherical shell. This shell encases a ball on which are marked the horizon, altitude lines, azimuths, etc. Several small holes in the ball permit moving the pivot to adjust the instrument to the proper latitude. (Figure 16)

1. Maddison (1962).

2. See Rico y Sinobas (1863-68).

3. See Maddison (1966).

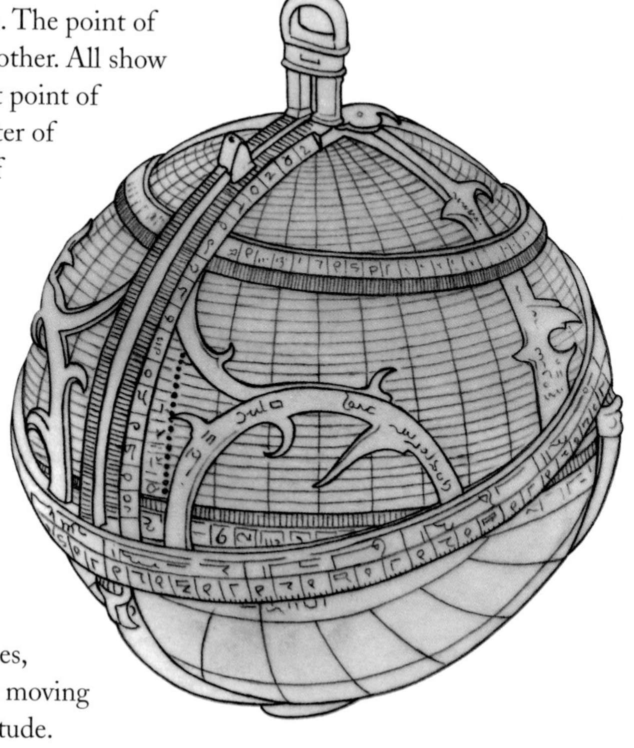

FIGURE 16 *The Spherical Astrolabe. Webster (1984), 3.*

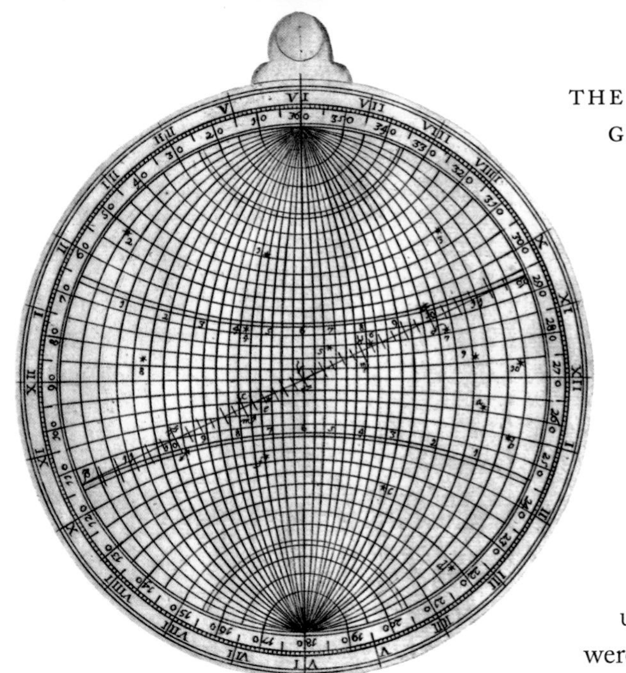

FIGURE 17 *The Planispheric Astrolabe of Azarquiel and Gemma Frisius. Regnartius (1610), at end of book.*

THE PLANISPHERIC ASTROLABE OF AZARQUIEL AND GEMMA FRISIUS In a planispheric astrolabe, the features on the surface of a sphere have been projected onto a flat surface or plane. In the case of the Azarquiel astrolabe, the centers of projection are the points of the vernal and autumnal equinoxes and the plane is the one that passes through the solstices. It is a stereographic projection that shows both the meridians and parallels as arcs of circles.

Azarquiel (fl. 1078-1100) described this projection in Toledo in the eleventh century. Reinvented by Gemma Frisius (1508-1555), the great mathematician at the University of Louvain, sometime before 1555, it was described by him as the Astrolabium Catholicum, or the universal astrolabe. Many beautiful examples of this astrolabe were made in Louvain by Gemma's nephews, the Arsenius brothers. (Figure 17)

THE DE ROJAS ASTROLABE This instrument was described by Joanne de Rojas, in 1550. De Rojas was a pupil of Gemma Frisius, and his astrolabe had some similarities to the Astrolabium Catholicum. He also used the plane that passes through the solstices, but he pulled back the center of projection to an infinitely distant point on the line passing through the equinoxes. It is thus properly described as an orthographic projection, and it is easily recognized, as the parallels of the celestial sphere become straight lines parallel to the equator. (Figure 18)

Although this astrolabe is known as the de Rojas instrument, examples exist that antedate his publication, notably the astrolabe made for Martin Bylica by the instrument maker Hans Dorn in the late fifteenth century.[3]

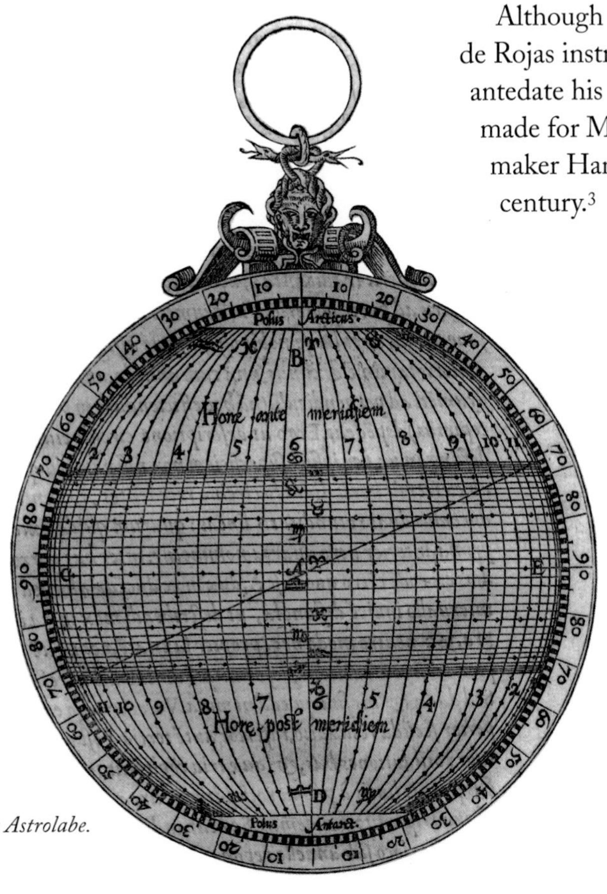

FIGURE 18 *The de Rojas Astrolabe. De Rojas (1551), 247.*

THE LA HIRE ASTROLABE The astrolabe of Philippe de la Hire (1640-1718) is described by Nicolas Bion (*c.* 1652-1733) in his handbook, *L'usage des astrolabes*, of 1702. The projection used is very similar to that of Azarquiel, the only difference being in the choice of the center used. La Hire placed the projection center at a point somewhere between the vernal equinox and infinity. The result was a figure in which the meridians and parallels were shown as arcs of ellipses and spaced at equal distances. This system was used by Bion on one of the plates in the cardboard astrolabe that he issued with his handbook. (Figure 19) Few other examples of the la Hire astrolabe are known, and it may be that Bion was angling for the very nice commendation he received for his work from the *Lecteur et Professeur Royal en Mathématique et de l'Academie Royale des Sciences,* Philippe de la Hire.

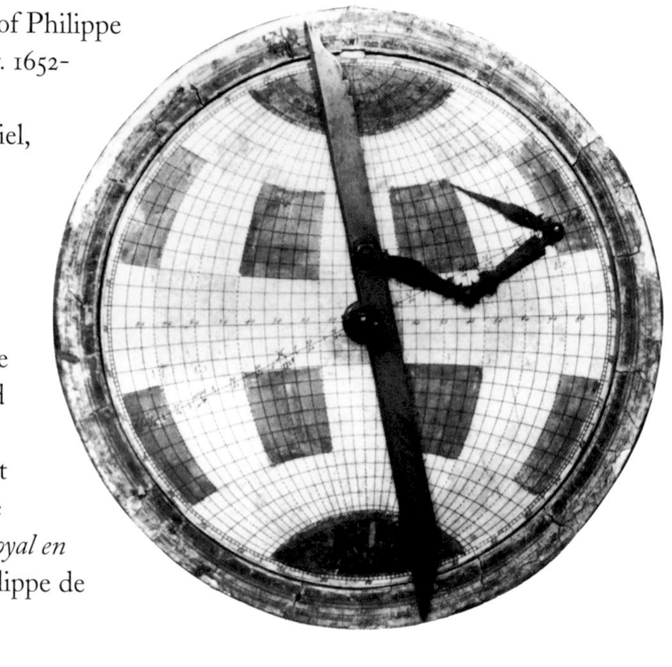

FIGURE 19 *The la Hire Astrolabe. Cardboard astrolabe by Nicolas Bion, National Maritime Museum, Greenwich.*

THE MATHEMATICAL JEWEL OF JOHN BLAGRAVE In 1585 John Blagrave (*c.* 1558-1612) published a book entitled *The Mathematical Jewel.* In it he described a planispheric astrolabe that was a composite of several earlier instruments. (Figure 20) It drew heavily on the work

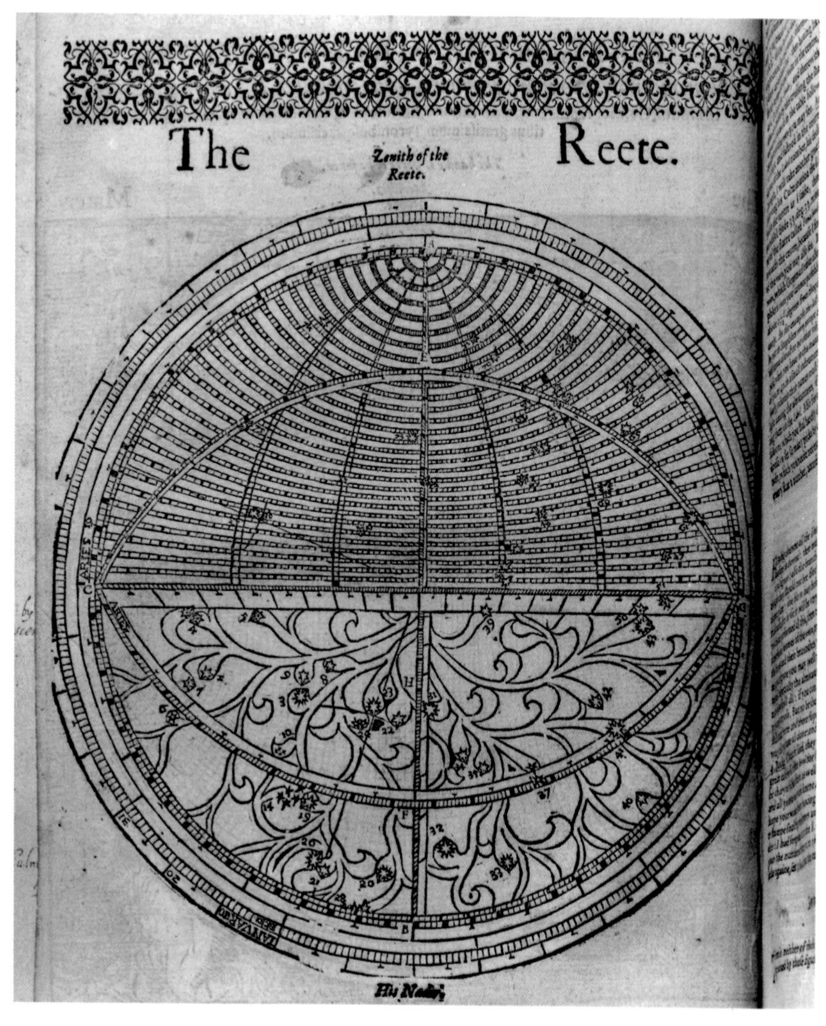

FIGURE 20 *The Blagrave Astrolabe. Blagrave* (1585), *between 16 and 17.*

4. King (1987a).

5. Blundeville (1594); Palmer (1658).

6. Waters (1966); Stimson (1988).

of Gemma Frisius and his Astrolabium Catholicum, but it also incorporated the Meteoroscopion of Petrus Apianus (1495-1552), which was described in his *Astronomicum caesareum* (1540). Although Blagrave claimed this as a new invention, it seems to have been used in a fourteenth-century astrolabe in the Benaki Museum in Athens.[4] It is a difficult instrument to use; mental gymnastics are required to see that the meridian lines also serve as horizons. Besides Blagrave's book describing this instrument, two later books appeared that attempted to explain its use.[5]

THE MARINER'S ASTROLABE

This instrument is not an astrolabe at all, in the classic sense. It is a heavy bronze disc marked on the limb in degrees. At the center is a pivot about which the alidade, or sighting bar, can rotate. The whole is supported by a ring and shackle so designed that the instrument will always hang vertically. (Figure 21)

FIGURE 21 *Mariner's Astrolabe, A-275 (cat. no. 46).*

Designed for use at sea, it was deliberately made heavy so that its inertia would help stabilize it. In addition, most mariner's astrolabes were cast in the form of a spoked wheel to minimize the effect of the wind upon them. This was a sighting instrument only and was used to measure either the altitude of the sun or a star or their zenith distance. The mariner's astrolabe first appeared in the fifteenth century and was superseded by more accurate instruments within the next 200 years.[6]

1 Classic-type astrolabe

England
c. 1250
Unsigned
29.0 x 20.9 x 2.61 cm.
Rete thickness — 0.4 cm.
Brass
ICA 200
M-26

This is an extremely early astrolabe, one of a small group that have zoomorphic retes, with quatrefoil decorations, that are thought to be among the first to have been made in Europe.

The mater consists of a cast rim with the backplate riveted to it. The throne takes the form of a demi-quatrefoil and has the original bail and suspension ring.

The face has a degree scale, marked and labeled every five degrees, 0° to 360°, from the top clockwise. The cavity is plain except for the slot for the tangs of the tympans. A bolt with a rectangular head is let into the rim at the 180° point. It has a hand-cut thread and a contemporary wing nut.

The backplate shows the conventional double calendar scales, which are eccentric. All the letters and numerals are in Gothic script and the names are in Latin. The zodiacal names are given.

Inside the calendars, the next three circles show a list of saints' days with their dates and a table of Golden Letters (■).

Next are eight concentric half-circles above the center line. From outside in, going counter-clockwise, they are:

1) "CI SOL^{9}" (circle of the sun), followed by 28 numbered spaces.

2) ",´CR^{9}" (mansions), followed by 28 numbered spaces: 1, 2, 3, 5, 6, 7, 1, 3, 4, 5, 6, 1, 2, 3, 4, 6, 7, 1, 1, 4, 5, 6, 7, 2, 3, 4, 5, 7.

3) "AN' BIS^{9}" (leap years), followed by "B'" in every fourth space starting under the 4 in the first row, then under 8, 12, 16, etc., through 28.

4) "LIT DOR" (Golden Letters), followed by 28 letters: F, E, D, B, A, G, F, D, C, B, A, F, E, D, C, A, G, F, E, C, B, A, G, E, D, C, B, G.

5) Blank

6) "CI T^{9}" (lunar cycle), followed by 19 spaces, numbered 1 to 19.

7) "CLAV^{9}," followed by 19 spaces, numbered 16, 15, 34, 23, 12, 31, 20, 39, 28, 17, 36, 25, 14, 33, 22, 11, 30, 19, 38.

8) "EPACTE" (Epacts), followed by 18 numbered spaces: 11, 22, 3, 14, r5, 6, 17, 28, 9, 20, 1, 12, 23, 4, 15, 26, 7, 18. The fifth space should be 25.

The shadow square is divided into twelve parts and labeled every two parts. Superimposed on the shadow square, and originating from the center point, is a double unequal hour scale.

The rete is arranged in a quatrefoil and demiquatrefoil pattern with 41 star points, some of which are zoomorphic (★). The back of the rete shows construction lines, with the ecliptic circle divided by degrees.

The "I" in BETIN CAITOZ, on the rete, is inserted horizontally over the

★ STAR LIST

Star name	*Modern name*
SCEDER	Alpha Cassiopeiae
MIRAC	Beta Andromedae
HV EQVI [erased]	Beta Pegasi
SCHCACK	Delta Aquarii
MV EQI	Epsilon Pegasi
DHENEB ALGEDI	Delta Capricorni
DHERAT	Alpha Cephei
ADDIGEGE	Alpha Cygni
DELFIN	Epsilon Delphini
ALTAIR	Alpha Aquilae
WEGA	Alpha Lyrae
TABEN	Gamma Draconis
ALHAWE	Alpha Ophiuichi
CORVS^{9}	Alpha Scorpii
YED	Delta Ophiuchi
ELFECA	Alpha Coronae Borealis
ALRAMEC	Alpha Bootis
BENCENAZ	Eta Ursae Majoris
ASSIMECH	Alpha Virginis
ALGORAB	Gamma Corvi
CAVDA	Beta Leonis
DECORVO	? Corvi
DVBHE	Alpha Ursae Majoris
SVBPEDE	Theta Ursae Majoris
COR	Alpha Leonis
ALFARD	Alpha Hydrae
MARKEB	Chi Velorum[1]
ALGOMERZA	Alpha Canis Minoris
RAZ	Alpha Geminorum
ALHABOR	Alpha Canis Majoris
ELGEVZE	Alpha Orionis
RIGEL	Beta Orionis
ALHAIOC	Alpha Aurigae
ALDEBARN	Alpha Tauri
AVGETAVAR	Tau Eridani
ALGENB	Alpha Persei
MENKHAR	Alpha Ceti
ENF	Alpha Arietis
BETIN CAITOZ	Zeta Ceti
ALFRAZ	Alpha Andromedae
DHENEB CAITOZ	Beta Ceti

The symbol "9" is the Gothic shorthand for the ending "us."

Face of No. 1

Reverse of No. 1

SAINTS' DAYS

Saint's day	*Date*	*Golden Letter*
IANIHVARIVS		
CIRC^{9}:	I	A
EPI:	6	F
MARCL^{9}:	16	B
S. PAUL^{9}:	25	D
FEBERHVARI^{9}		
BLAS^{9}:	3	F
CAT^{9}PE:	22	D
MATH^{9}	24	F
MARCIVS		
ĀNONTIA T̊	25	G
APRILLIS		
MARCHI	25	G
MAIVS		
VRBANI	25	E
IVNIVS		
BARNABE	II	A
ION^{9}IS	24	G
IVLIVS		
MARTIN^{9}:	4	C
MAGDAEN^{9}	22	G
AVGVSTVS		
LAVREN^{9}:	10	E
BARTHOL^{9}	24	E
SEPTMBER		
NATMAR^{9}:	8	F
NICHAEL^{9}	29	F
OCTOBER		
DION^{9}:	9	B
LVC^{9}:	18	D
SVM^{9}. IVD^{9}	24	G
NOVEMBER		
OMN^{9}:	I	D
S.^{9}MART̄	II	GG
NAT^{9}AD	30	E
DECEMBER		
nICHOL^{9};	6	D
THOM^{9}:	20	E
NATAL^{9}	25	B

The symbol "9" is the Gothic shorthand for the ending "us."

Rete of No. 1

"N" as an afterthought. The "N"s are formed by two "I"s with a diagonal slash connecting them, sometimes one way, sometimes the other. "AL" and "AT" are sometimes elided.

There are five tanged, silvered tympans, all labeled "LATITUDO," for 40° and 42° 45', 43° 30' and 60°, 47° 47' and 51°, 48° 48' and 52° (the 48° 48' side is labeled "PARISIUS," and the unequal hour lines for this tympan, labeled I to VI, were added later by a different hand), and 44° and 50° (the 44° side has 1 to 24 added later in modern Arabic numerals, around the outer edge). In addition to the usual lines, the Great Houses and unequal hours are shown, the latter labeled with Gothic numerals. The almucantars are drawn every three degrees. The azimuth lines are drawn and labeled every ten degrees.

The alidade and rule are replacements.

Gunther described this astrolabe under two entries, nos. 200 and 295.[2] The ICA picked no. 200, although no. 295 has the more accurate description.[3] The retes of ICA 293 and 294 are very similar to that of M-26.[4] There are at least four other early English astrolabes in this group, all inscribed in Latin and engraved in Lombardic capitals. See the excellent entry by F. R. Maddison[5] and the article by Gingerich.[6]

DATE ACCORDING TO PRECESSION — *c.* 1239.
I ARIES = 12.5 March, from the calendar scales.
PROVENANCE — A. W. M. Mensing, Amsterdam, 1924; Max Adler, Chicago, 1930; A. P. gift, 1930.
EXHIBITION — *The Secular Spirit: Life and Art at the End of the Middle Ages,* Metropolitan Museum of Art, The Cloisters, New York, March 26 - June 3, 1975, item 1015.
REFERENCES — Engelmann Catalogue (1924), 14, plates 1, 6; Gibbs *et al.* (1973), 14; Gingerich (1987), 95, 98-99, 103; Gunther (1932), 348, 472, plate 129, fig. 185; Hayward Catalogue (1975), 180 with picture; Price (1955), 250; Sotheby's (1986), 24, lot 125.

1. Kunitzsch (1959), 177.
2. Gunther (1932), 348, 472, plate 129.
3. Gibbs *et al.* (1973), 14.
4. Gunther (1932), 469-71, plate 118, fig. 185.
5. Sotheby's (1986), 24, lot 125.
6. Gingerich (1987), 95, 98, 103, table 1.

2 Classic-type astrolabe

[Johanne Fusoris]
Paris
c. 1400
Unsigned
20.6 x 16.3 x 3.3 cm.
Rete thickness — 0.8 cm.
Brass
ICA 199
M-27

Face of No. 2

This astrolabe is an outstanding example of a working instrument for a medieval astronomer. Although unsigned, it is certainly the work of Johanne Fusoris, one of the best-documented of the mathematicians and instrument makers of Paris in the fifteenth century. All the numerals and letters are in a Gothic script.

The mater consists of a cast rim with the back soldered to it. The throne is minimal: a round boss with small lugs on either side, cast as part of the rim. The boss is drilled for a pin that holds the bail and suspension ring. The face of the mater shows an hour scale labeled 12 to 12 twice, divided by degrees, and marked every five degrees or 20 minutes. There is an old, neat repair at four o'clock on the right side. The cavity is blank except for the slot for the tangs on the tympans.

The back of the astrolabe shows an eccentric double calendar (■). Within the calendar is a double unequal hour scale labeled 1-6-1 and a shadow square divided into twelve parts and labeled every four parts.

The rete carries 21 named star points (★). Elfeta is broken off and Cornu (Beta Arietis) is misplaced, as it is on all Fusoris astrolabes. The reverse of the rete shows the construction lines used by the maker. Virgo shows as "Virg" on the ecliptic circle.

There are four tympans, drawn for the latitudes 42°, 45°, 49°, and 52°. The almucantars are shown every two degrees, numbered through ten degrees on all the tympans and dotted for every multiple of ten degrees on the 42° and 45° plates. The azimuths are drawn every ten degrees. All of the tympans show an unequal hour diagram labeled every other hour. The 49° plate, which is labeled "PARICI⁹," is the only one to have a diagram on the reverse. It shows eighteen horizons. All of the tympans show the meridian, the right horizon, the Tropics, and the equator on both sides.

The alidade and the double rule are counterchanged, with the ends

★ STAR LIST

Star name	Modern name
humer^{9}	Beta Pegasi
ariof	Alpha Cygni
althair	Alpha Aquilae
wegua	Alpha Lyrae
alhave	Alpha Ophiuchi
cor	Alpha Scorpii
alcameth	Alpha Bootis
elfeta [broken off]	Alpha Coronae Borealis
eq^{9}	Eta Ursae Majoris
spica	Alpha Virginis
cor le	Alpha Leonis
ydra	Alpha Hydrae
algimeica	Alpha Canis Minoris
alhabor	Alpha Canis Majoris
algerice	Alpha Orionis
rigel	Beta Orionis
alhatot	Alpha Aurigae
aldebaran	Alpha Tauri
memkar	Alpha Ceti
cornu	Beta Arietis
venter ceti	Zeta Ceti

The symbol "9" is the Gothic shorthand for the ending "us."

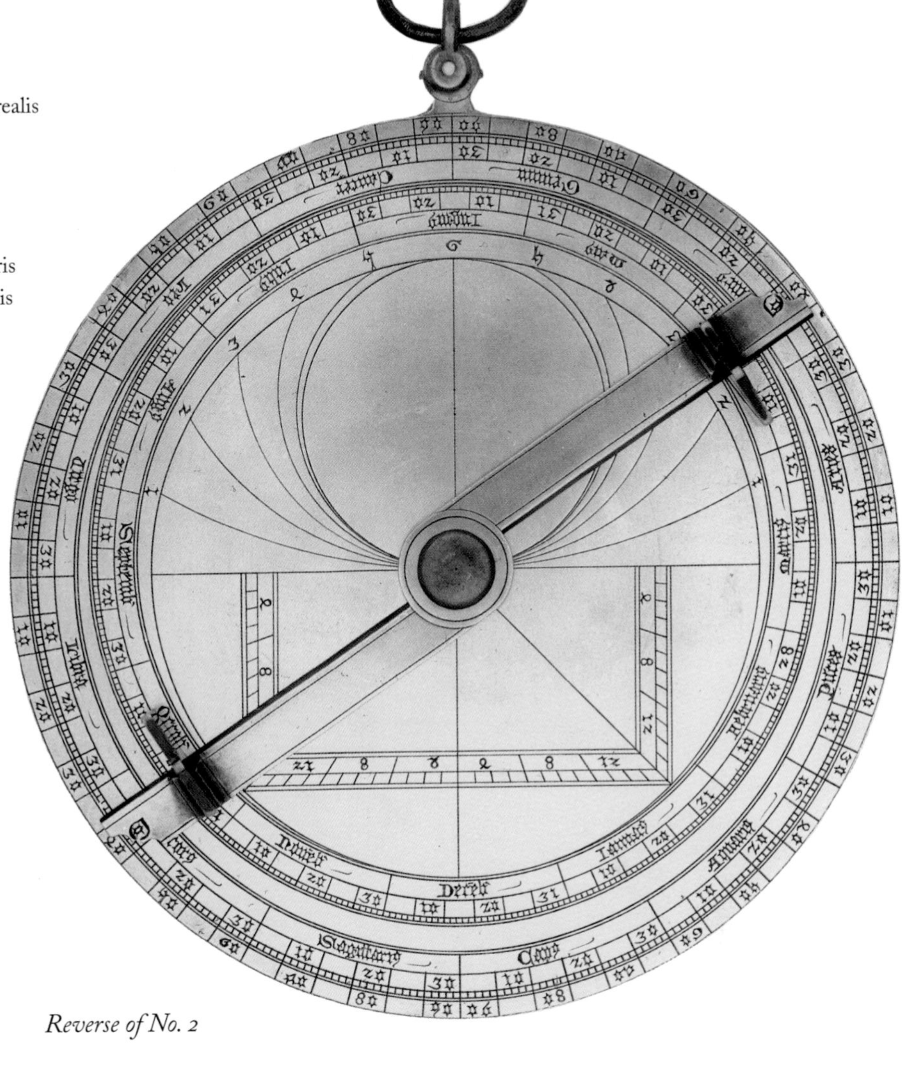

Reverse of No. 2

having Gothic, cut-out decoration. The bolt and horse survive intact.

M-27 is one of a group of medieval astrolabes clearly by the same maker. The calligraphy, the throne, the rete, and the misplaced star (Cornu-Beta Arietis) are almost identical in all the examples. Emmanuel Poulle[1] has clearly demonstrated that the maker of these astrolabes was Johanne Fusoris. Poulle located several manuscripts written and illustrated by Fusoris and used by him for his defense during his trial for treason in 1416.

DATE ACCORDING TO PRECESSION — *c.* 1400.

1 ARIES = 11.5 March, from the calendar scales.

PROVENANCE — A. W. M. Mensing, Amsterdam, 1924; Max Adler, Chicago, 1930; A. P. gift, 1930.

EXHIBITION — *Tokens of Possession,* Royal Ontario Museum, Toronto, Feb. 2 - March 28, 1976.

REFERENCES — Adler (1970), 17; Adler (1973), 23; Adler (1980), 24; Engelmann Catalogue (1924), 14, plates 1, 6; Gibbs *et al.* (1973), 14; Gunther (1932), 348, plate 83; Murdoch (1984), 264; Poulle (1963), 20, 22-23; Price (1955), 251.

1. Poulle (1963).

■ DOUBLE CALENDAR

Month name	Zodiac name
Marci^{9}	Aries
Aprilis	Taur^{9}
Mai^{9}	Gemin
Iuni^{9}	Cancer
Iuli^{9}	Leo
Aug^{9}	Virgo
Septemb	Libra
Octob	Scor^{9}
Noveb	Sagittari^{9}
Deceb	Cap^{9}
Ianvari^{9}	Aquari^{9}
Febrvari^{9}	Pisces

3 Classic-type astrolabe

[Johanne Fusoris]
Paris
c. 1400
Unsigned
20.6 x 16.3 x 1.1 cm.
Rete thickness — 0.8 cm.
Brass
ICA 193
W-264

Face of No. 3

This astrolabe is almost identical to M-27 (cat. no. 2), which is by the same maker. We note the following differences on W-264:

1) The rete is identical except that Algerise is not labeled and the pointer for Elfeta is intact.

2) The four existing tympans are drawn for 42° and 36°, 45°, 49°, and 52°. The last three are plain on the back except for the Tropics, equator, meridian, and right horizons, all drawn by Fusoris. The 36° side was done in a later hand, and the numerals are not in Gothic style. The crepuscular line is shown. The 49° tympan is labeled "Paricius." Under the label, "49" is scratched in Gothic numerals. This was probably done by Fusoris to indicate which tympan it was. The almucantars are drawn every two degrees and numbered through ten degrees.

3) On the back, the months are the same as in M-27. The zodiac is the same except for "Sagita" instead of "Sagittari⁹."

4) The rule is single with a simple outline as decoration. The ends are simple points.

There are four vellum rings to replace missing tympans. These are each labeled: "Cette pièce en cuivre / appartient à m. Mercier / marchand

Reverse of No. 3

horloger à / Thiviers." Mercier was working in Thiviers as a clockmaker from 1834 to 1840.[1]

DATE ACCORDING TO PRECESSION — *c.* 1400.
I ARIES = 11.5 March, from the calendar scales.
PROVENANCE — M. Mercier, Thiviers, 1834-40; Sir John Findlay, Edinburgh, 1930; Harriet, Lady Findlay, Edinburgh, 1930; Sir Edmund Findlay, Edinburgh, 1954;[2] on loan to the Royal Scottish Museum, 1930-61; Nicolas Landau, Paris, 1961; R. S. Webster, Winnetka, Ill., 1964; A. P. gift, 1991.

REFERENCES — Adler (1992), cover and inside cover page; *Curator* (1995), cover; Gunther (1932), 342-43; Poulle (1963), 20, 22-23; Sotheby's (1961), 33, lot 136, plates 15, 16; Turner, A. J. (1985), 38, fig. 28.

1. Tardy (1972), part 2: 457.
2. Morrison-Low correspondence, 1993.

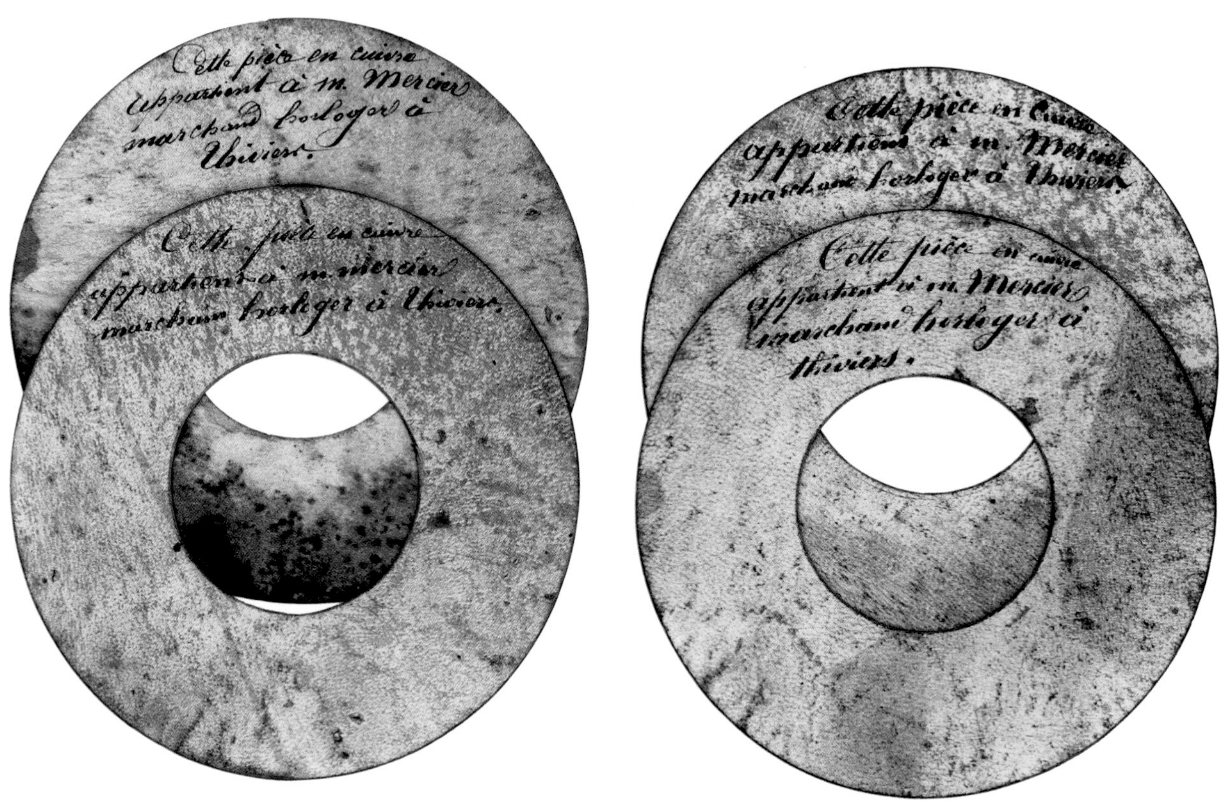

ABOVE: *Vellum spacers of No. 3*
BELOW: *Rete of No. 3*

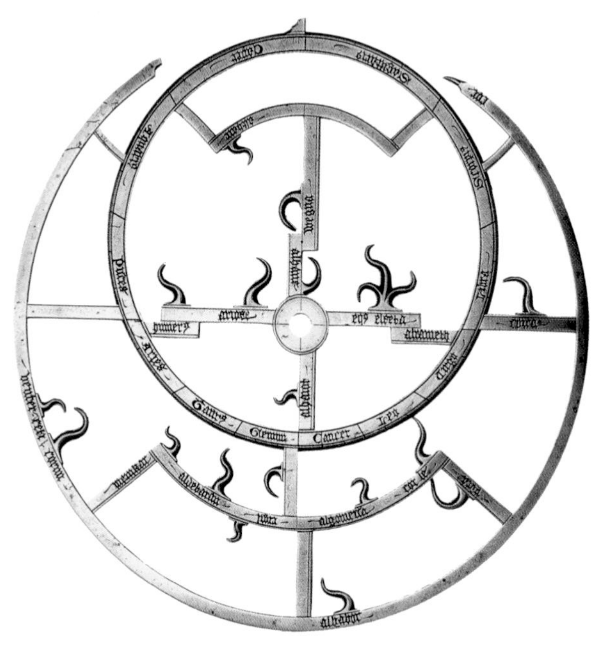

4 Classic-type astrolabe

Spain
c. 1500?
Unsigned
41.0 x 32.2 x 4.3 cm.
Rete thickness — 0.16 cm.
Brass
ICA 164
M-28

Face of No. 4

The mater is formed from a cast ring with the backplate riveted to it. The arched throne is riveted to a hoop, which is in the form of a dragon swallowing its tail. This hoop lies in a trough or gutter recessed into the rim of the mater. The face of the mater is divided into 360 degrees, labeled every five degrees, 0° to 360°, and by hours, 1 to 24, from the bottom clockwise.

The bottom of the cavity is divided by degrees, labeled every five degrees, 0° to 60°, six times from the top counter-clockwise. The 60° points are labeled 1 to 6. The degree marks are connected vertically, forming a sine table. The horizontal center line is flanked by two flattened double arcs, numbered 1 to 9 from the meridian. The center line is labeled "EQVATIONES CIRCVLI," with the additional notation "MI NVS" and "AS DE" on both sides of the central hole.

The reverse of the mater shows an interesting variant of the double calendar. It is made up of four rings, three fixed and one movable. The outer, zodiacal ring is riveted to the backplate, with "CANCER" at the top. It has two touch points to locate the first points of Aries and Libra. The names of the signs are given. Next is a fixed eccentric ring, divided by degrees and labeled "AVS" at the top,

Reverse of No. 4

narrow point. Within this is the movable civil calendar ring with the names of the months in Latin. The fourth ring, which is eccentric and unmarked, serves as a guide for the adjustable ring. This arrangement makes it possible to update the sun's position in the zodiac.

Inside the calendars, a circle of degrees is inscribed on the backplate. It is labeled 90°-0°-90° twice from the top. A double unequal hour scale is numbered 1 to 12 from both sides and labeled "ASCENSVS" and "DESCENSVS" on each side. It is labeled "LINAE MERIDIES" twice within the hour scale.

The shadow square is divided into 24 half-parts, marked and labeled every part, 1 to 12, and also labeled "VMBRA VERSA" and "VMBRA EXTENSA." The 45° lines from the center to the corners are labeled "GNOMO."

The rete extends beyond the Tropic of Capricorn and is composed of 60 small pieces of brass, which have been hammered, shaped, filed, riveted, and soldered together. It shows the positions of 35 stars (★). There are several zoomorphic elements, including a hand with the index finger extended, which is set at the top of the ecliptic circle. It is

★ STAR LIST

Star name	*Modern name*
PECT' CASIEPIE	Alpha Cassiopeiae
HVMERVS PEGASI	Eta Pegasi
CRVS MERIDIONALE	Alpha Crucis
MVSIDA EQVI PEGASI	Epsilon Pegasi
CAVDA GALINE	Alpha Cygni
AQVILA VOLANS	Alpha Aquilae
AQVILA CADENS	? Lyrae
CAPVT SERPENTARII	Gamma Draconis
COR SCORPII	Alpha Scorpii
CORONA	Alpha Coronae Borealis
ARCTVRVS	Alpha Bootis
IN THEMONE PRIMA	Eta Ursae Majoris
SPICA	Alpha Virginis
CORVVS	? Corvi
CAVDA LEONIS	Beta Leonis
IN VRSA MAIORE	Alpha Ursae Majoris
COR LEONIS	Alpha Leonis
EQVVS PRIOR	Alpha Equulii
NAVIS	? Argo Navis
IN COLLO CAĪS	Beta Canis Minoris
CAPVT PROMIGET NOR	? Geminorum
IN ORE CANICVLE SIDVS MAGNE LVCIS	Alpha Canis Majoris
HVS ORIONIS	Alpha Orionis
CAPVT CANIS	? Canis
PES ORIONIS	Beta Orionis
AVRIGA	? Aurigae
OCVLVS TAVRI	Alpha Tauri
ERIDANVS	? Eridani
TAIGETE PLEIAS	Eta Tauri
LAT^{9} DEXTRV̄ PSEI	Alpha Persei
NARIS CETI	Alpha Ceti
FLVXVS	Alpha Pisces
VENTER CETI	Zeta Ceti
CAP̄ ANDROMEDE	Alpha Andromedae
CAVDA CETI	Beta Ceti

The symbol "9" is the Gothic shorthand for the ending "us."

Mater of No. 4

labeled "OSTENSOR" on the rim of the rete. The ecliptic circle is drawn showing all the divisions radiating from the pole of the ecliptic.

There are four tympans, each with notches top and bottom, to fit the lugs in the mater. Three carry the usual lines plus the Great Houses and unequal hour scales below the horizon. The latter are labeled 1 to 12 on the right side. These three tympans are drawn for latitudes 40° and 42°, 44° and 45°, and 48° and 50°. The almucantars are shown every two degrees and the azimuths every ten degrees.

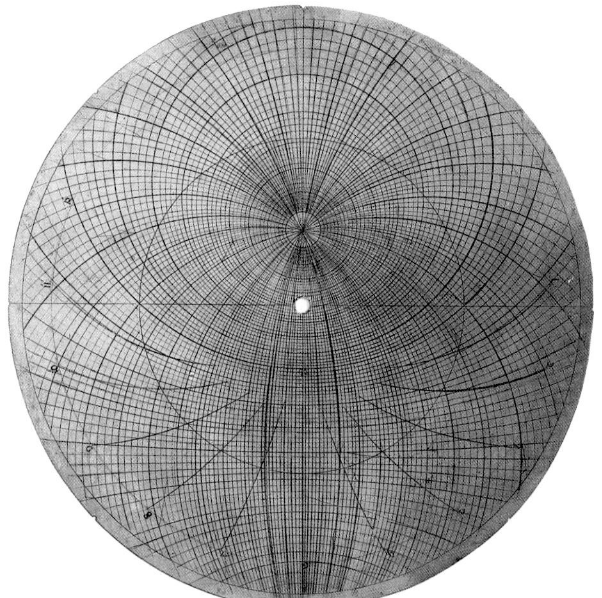

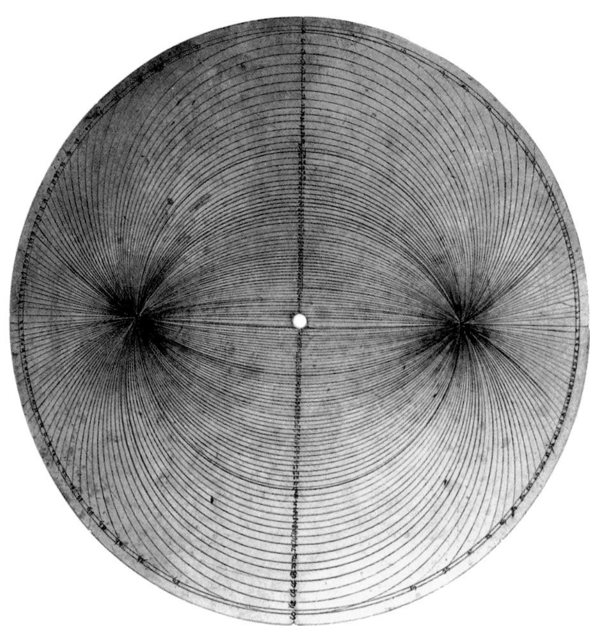

LEFT: *Tympan #3 (for 50°) of No. 4*
RIGHT: *Tympan #4 (multiple horizons) of No. 4*

The fourth tympan has multiple horizons on one side and is drawn for the latitude of 66° 30' on the other. The signs of the zodiac are inscribed around the circumference. The almucantars within the ecliptic are shown every degree.

The alidade has four counter-changed arms and four pinhole sights. It is marked "ASCENSVS" and "DESCENSVS" and shows numbers from 1 to 12, for use with the unequal hour scale. The counterchanged double rule is plain.

The script and numerals, except for the Gothic fives, are not typical of a late fifteenth-century instrument.

This very interesting astrolabe, although mathematically correct, was not made by a professional instrument maker, and it remains an enigma.

DATE ACCORDING TO PRECESSION — *c.* 1498.
1 ARIES = 21 March, from the calendar scales, as currently set.
PROVENANCE — A. W. M. Mensing, Amsterdam, 1924; Max Adler, Chicago, 1930; A. P. gift, 1930.
EXHIBITIONS — *Convivencia: Jews, Muslims and Christians in Medieval Spain,* The Jewish Museum, New York, Sept. 20 - Dec. 20, 1992; The Meadow Museum, Dallas, Jan. 29 - April 11, 1993.
REFERENCES — Engelmann Catalogue (1924), 14, plates 1, 6; Gibbs *et al.* (1973), 13; Gunther (1932), 311, plate 70; Mann *et al.* Catalogue (1992), 88, 242; Price (1955), 251.

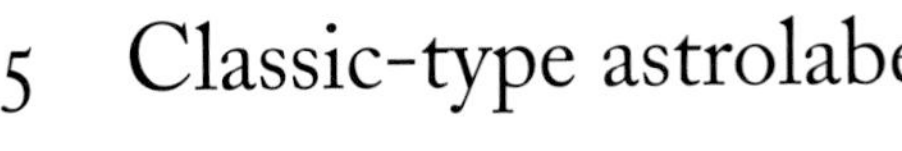

5 Classic-type astrolabe

Georg Hartmann
Nuremberg
1532
"GEORGIVS HARTMAN / NORENBERGE FACIEBAT / ANNO M D XXXII"
in the shadow square
19.0 x 13.7 x 2.6 cm.
Rete thickness — 0.6 cm.
Brass
ICA 4549 New Series [initiated by David A. King]
W-272

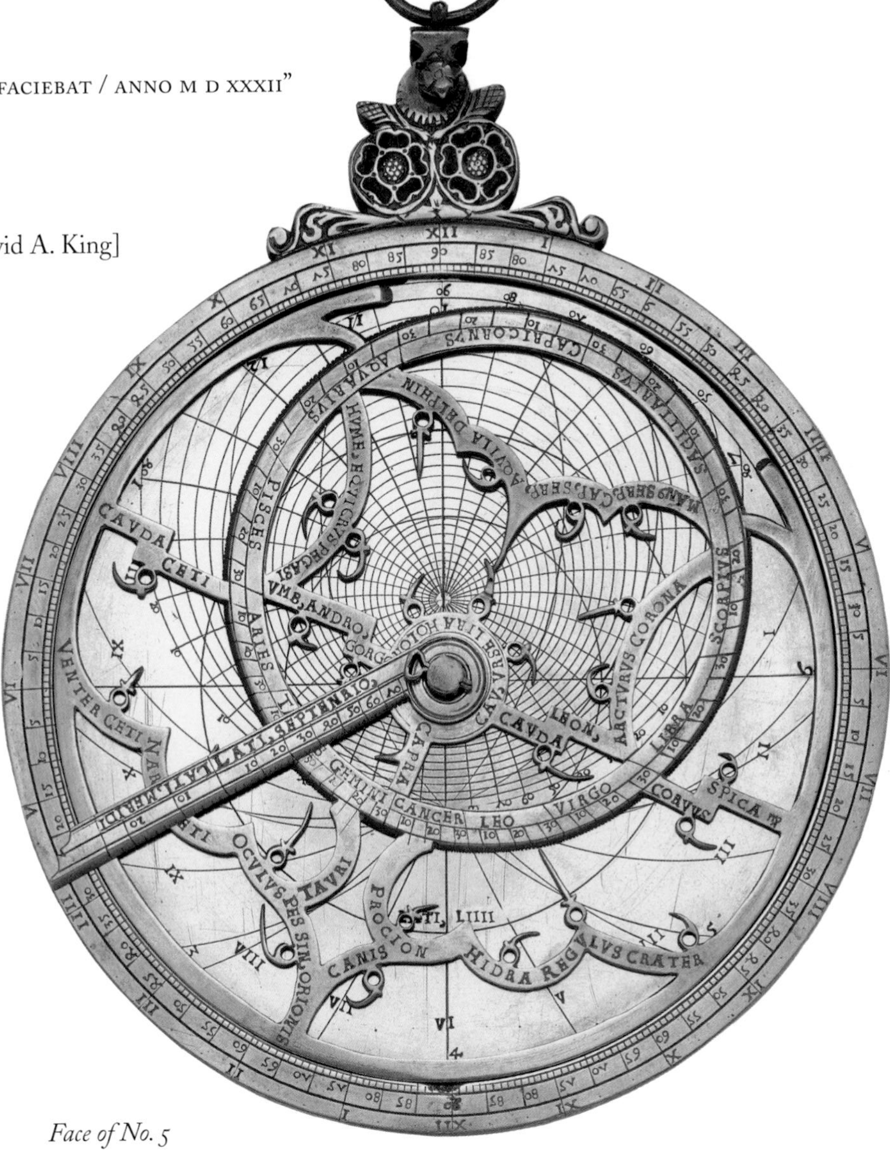

Face of No. 5

The mater consists of a cast ring soldered to the hammered backplate. The throne was cast separately and soldered to the rim. There are two holes near the ends of the throne that extend through the rim; these were used to align the throne on the rim, prior to soldering. It carries the usual two rosettes adopted by Hartmann. The ring and bail are present. The shackle may be a replacement.

The limb has an hour scale labeled XII to XII twice, combined with a degree scale, which is divided every degree and marked and labeled every five degrees, 90°-0°-90°, twice. There are two later rivets that interrupt the numerals on the ends of the horizon line. The Arabic numerals are Gothic, and all the numerals on the face are stamped.

The back of the astrolabe carries a scale of degrees on the limb, divided every degree and marked and labeled every five degrees, 90°-0°-90°, twice. Next is the zodiacal calendar, labeled with Latin names and symbols. It is also marked and labeled X, XX, XXX twelve times. The civil calendar is divided into days, marked and labeled by fives, 5 to 28, 30, or 31 as appropriate. The months are in Latin. Both calendars are concentric.

Above the center line there is an unequal hour diagram, engraved 1 to 12 in Gothic numerals. The usual shadow square below is divided into twelve parts and labeled by threes. It is also labeled "VMBRA VERSA" and "VMBRA RECTA." The signature lies within the shadow square. There is a later, coarsely engraved "22" below the square, probably an owner's mark.

The rete is a typical Hartmann-type, with the star-pointers springing from pierced bases. The ecliptic circle is divided as usual, with the zodiacal names punched in. Twenty-seven stars are shown (★).

The back of the rete shows the degree lines on the ecliptic circle and the right horizon.

COMPOSITION

Element	Weight %
Antimony	0.04994
Arsenic	0.00124
Bismuth	0.00000
Cadmium	0.00477
Cobalt	0.00000
Copper	81.24152
Gold	0.00000
Iron	0.62850
Lead	0.63901
Manganese	0.00000
Mercury	0.02976
Nickel	0.04083
Silver	0.03767
Tin	0.11695
Zinc	17.20982

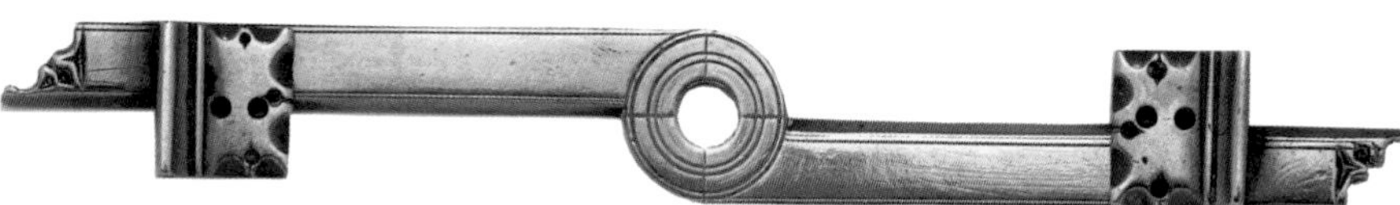

TOP: *Reverse of No. 5*
BOTTOM: *Alidade of No. 5*

The three tympans are for 39° and 42°, 45° and 48°, and 51° and 54°. They are labeled in Roman numerals, with 54° being LIIII. The latitudes are also engraved in Gothic numerals on the tangs. The Great Houses are labeled in Roman numerals and the unequal hours in Gothic. The azimuths are drawn and labeled every ten degrees, and the almucantars are drawn every three degrees, but not labeled. The three tympan faces and the back of the rete are lightly marked with the letter "i" or "l" near the lower edge; this is probably an assembly mark. The various parts of the Hartmann astrolabe at The Time Museum are marked with an "h/H."[1]

★ STAR LIST

Star name	*Modern name*
CRVS PEGASI	Beta Pegasi
HVME, EQVI	Alpha Pegasi
HOLOR	Alpha Cygni
DELPHIN	Alpha or Epsilon Delphini
AQVILA	Alpha Aquila
LIRA	Alpha Lyrae
CAP, SERP,	Beta Draconis
MAN^{9} SERP,	Delta Ophiuchi
CORONA	Alpha Coronae Borealis
ARCTVRVS	Alpha Bootis
CAV, VRSE	Alpha Ursae Majoris
SPICA ♍	Alpha Virginis
CORVVS	Alpha Corvi
CAVDA LEON,	Beta Leonis
CRATER	Alpha Crateris
REGVLVS	Alpha Leonis
HIDRA	Alpha Hydrae
PROCION	Alpha Canis Minoris
CANIS	Alpha Canis Majoris
PES SIN, ORIONIS	Beta Orionis
CAPRA	Alpha Capricorni
OCVLVS TAVRI	Alpha Tauri
NARES CETI	Alpha Ceti
GORGON	Beta Persei
VENTER CETI	Zeta Ceti
VMB, ANDRO,	Alpha Andromedae
CAVDA CETI	Beta Ceti

The symbol "9" is the Gothic shorthand for the ending "us."

Rete of No. 5

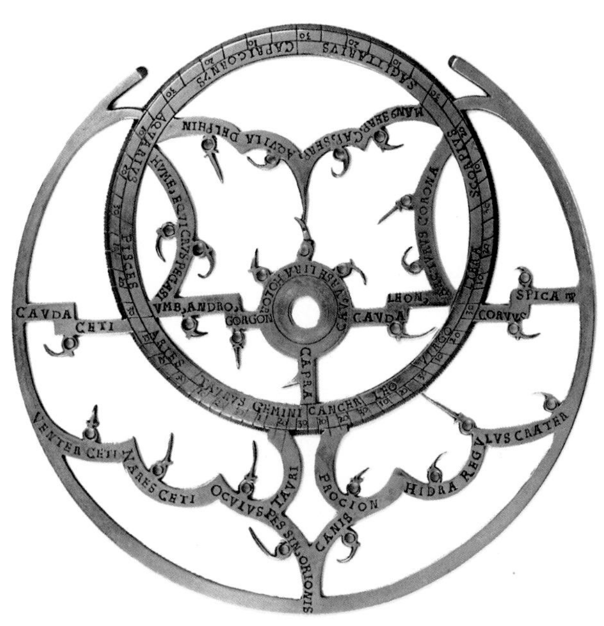

The alidade has folding sights with double pinholes. The single rule is divided along the beveled edge, marked 70°-0°-23.5°, and labeled every ten degrees, 70°-0°-20°, from the center to the tip. The inner section is labeled "LATI, SEPTENRIO" and the outer one "LATI, MERIDI." The construction lines for the declination scale show on the back of the rule.

The Winterthur Museum Analytical Laboratory tested the mater of this astrolabe in March 1988; the results are shown in the table (❊).

We know of five other Hartmann astrolabes made in 1532. They are in The Time Museum, Rockford, Illinois; the British Museum, London; the Germanisches Museum, Nuremberg; the Národní Technicki Muzeum, Prague; and the Musée de la Renaissance, Ecouen.[2]

1 ARIES = 10.8 March, from the calendar scales.
PROVENANCE — Christie's New York auction, 1985; London dealer, 1992; A. P. purchase, 1992.
REFERENCES — Bennett (1987), 15; Christie's New York (1985), 100-101, lot 334; Lamprey (1997), 111-42.

1. Lamprey (1997), 111-42.
2. Turner, A. J. (1985), 128-29.

6 Classic-type astrolabe

Georg Hartmann
Nuremberg
1540
"GEORGIVS HARTMAN / NOREMBERGE FACIEBAT / ANNO: M.D.XL"
under the shadow square
20.0 x 14.0 x 2.0 cm.
Rete thickness — 0.6 cm.
Brass
ICA 267
M-22

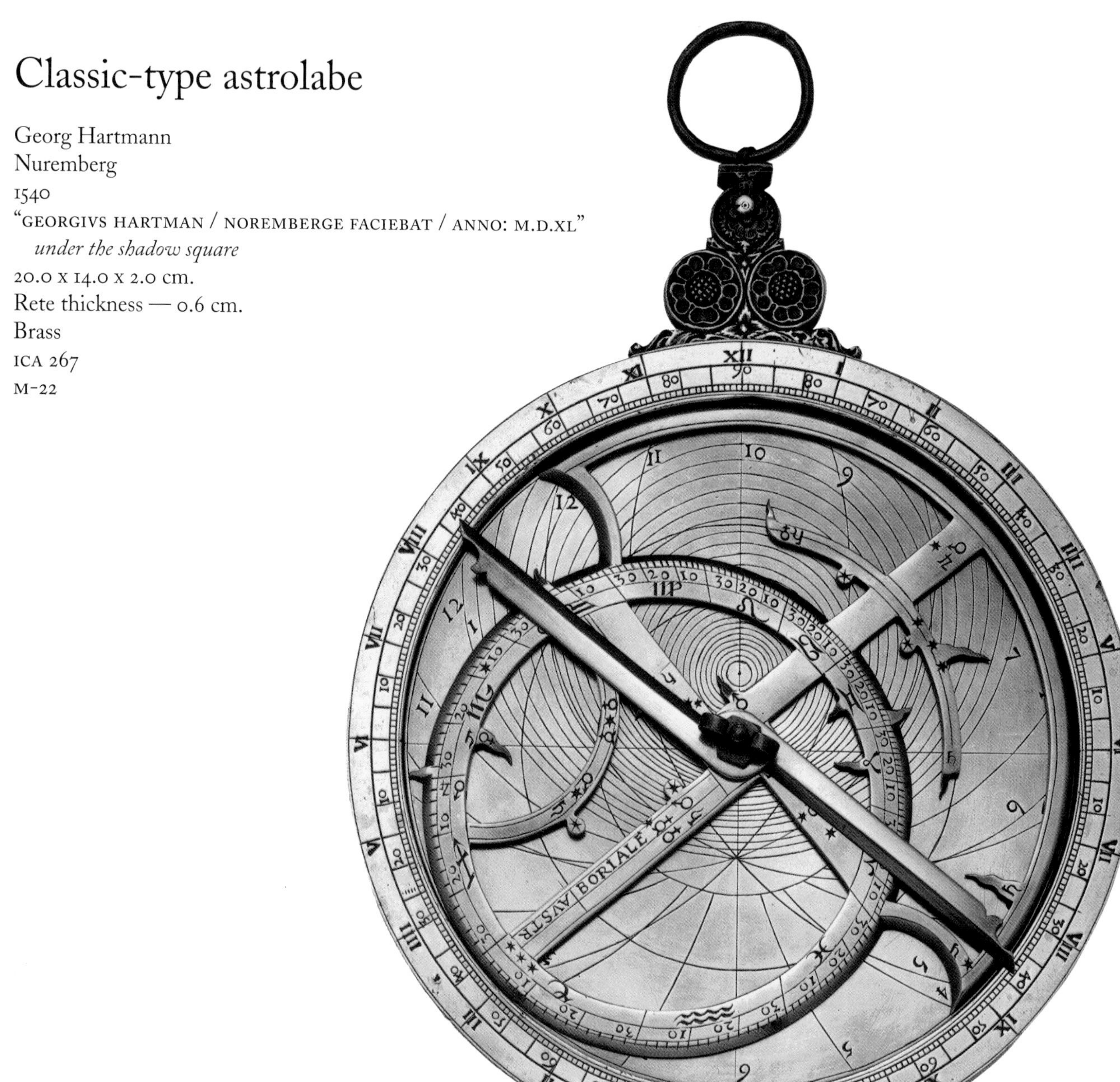

Face of No. 6

This Hartmann astrolabe has an atypical rete. The mater consists of a cast rim with the backplate soldered to it. The throne is decorated with three rosettes (a Hartmann trademark), and it is attached to the rim by brazing or soldering. The ring and swivel-bail are present.

The face of the mater has an hour scale on the limb, divided and labeled XII to XII twice. Inside is a circle of degrees marked every five degrees and labeled every ten degrees, 90°-0°-90°, twice from the top. The cavity shows the Tropics, the equator, the meridian, and the right horizon. There is a slot for the tangs and an assembly number, "2," stamped on the bottom near the rim. An assembly or production number is common in Hartmann astrolabes, although none of the other parts are so labeled in this example.

The back of the mater shows the conventional double calendar scales, which are concentric. The names of the months are in Latin, with "Marcius" being the only variant spelling. The names of the zodiac are given.

A double unequal hour diagram is shown above the center line. It is labeled 1 to 12 clockwise and surrounds a double-headed eagle. The shadow square below is divided into twelve parts and labeled 3, 6, 9, 12, etc. The square is also labeled "VMBRA VERSA" and "VMBRA RECTA" twice. The signature and date appear beneath it.

The rete shows 33 stars, some indicated by flame-shaped star points, others by asterisks. None are named but most are identified by their astrological symbols. The modern names of the stars shown are listed here in the order of their sidereal hour angle (★). This is not a typical Hartmann rete but appears to be identical to ICA 252.[1] It carries no assembly mark, and the thickness is not uniform. On the meridional bar a short section of the celestial equator is shown, labeled "AUSTR"

★ STAR LIST

Star sign	*Modern name*
*	Alpha Coronae Australis
*	Beta Coronae Australis
*	Gamma Coronae Australis
♀ ☿	Alpha Lyrae
♄ ♂	Gamma Draconis
♄ *	Alpha Ophiuchi
♂ *	Alpha Herculis
♃ ♂	Alpha Scorpii
♄ ♂	Delta Ophiuchi
♀ * ☿	Alpha Coronae Borealis
*	Alpha Librae
♄ * * * *	Ursae Majoris
♄ ☿	Gamma Bootis
♀	Alpha Virginis
♂	Alpha Ursae Majoris
♄ ♀	Alpha Hydrae
*	Alpha Canis Minoris
♃ ♂ *	Alpha Canis Majoris
*	? Zeta Orionis
*	Alpha Orionis
*	Epsilon Orionis
*	Delta Orionis
	[Beta Orionis]
	[Alpha Aurigae]
	[Beta Persei]
♄	Alpha Ceti
♄	Zeta Ceti
♄ *	Beta Ceti
♀ ♄ *	? Alpha Andromedae
♄ ♀ *	Alpha Cassiopeiae

Reverse of No. 6

outside the equatorial circle and "BORIALE" within it. The reverse of the rete is unmarked.

The single tympan is made from rolled brass, is 0.2 cm. thick, and appears to be a replacement with no assembly mark. The face is for 60° and shows the usual lines plus the Great Houses and unequal hours. The reverse shows nine horizons.

Both rules are modern replacements. The bolt and wing nut might be original.

Gunther's entry for M-22, ICA 267, cites ICA 257 and 263 as being similar.[2] In fact, it is ICA 252 that has an identical rete to M-22. Gunther called ICA 252 a German astrological astrolabe.[3] The Time Museum Catalogue incorrectly states that M-22 is made of wood and paper.[4]

DATE ACCORDING TO PRECESSION — *c.* 1460.
I ARIES = 11 March, from the calendar scales.
PROVENANCE — A. W. M. Mensing, Amsterdam, 1924; Max Adler, Chicago, 1930; A. P. gift, 1930.
REFERENCES — Engelmann Catalogue (1924), 14, plates 1, 6; Fox (1933), 35; Gibbs *et al.* (1973), 15; Gibbs with Saliba (1984), 150; Gunther (1932), 440, plate 109; Klemm (1990), 29, 80-81 (plate); Price (1955), 252; Turner, A. J. (1985), 43; Zinner (1965), 364.

1. Gunther (1932), 425, plate 105.
2. *Ibid.*, 438-40, plate 109.
3. *Ibid.*, 425, plate 106.
4. Turner, A. J. (1985), 43 n. 145.

7 Classic-type astrolabe

Europe
c. 1550
Unsigned
17.0 x 11.5 x 2.6 x cm.
Rete thickness — 0.65 cm.
Brass
ICA 159
M-20

Face of No. 7

This astrolabe is Gothic in style and inscribed in Hebrew. The throne is formed by a central boss, pierced by a trefoil cutout, and supported by two pierced brackets. The back of the throne has a three-petalled flower at the top and two rabbits, hiding in their burrows, beneath. The bail and ring are present. The rim and the throne are brass castings that are soldered to the backplate.

On the face, the limb carries Hebrew numerals marked and labeled every five degrees, 0° to 360°. The cavity shows the Tropics and the equator.

The backplate, of hammered brass, carries two eccentric calendars. The zodiacal calendar is labeled in Hebrew, and the Julian calendar shows the month names in Hebrew. The first point of Aries is at March 10.25. Within the calendars are a double unequal hour scale and a shadow square, divided into twelve parts and labeled 2 to 12 and 12 to 2 twice, also in Hebrew.

The rete carries 27 star-pointers, but only 20 are labeled (★). Stars nos. 1, 5, and 12 from Bernard Goldstein's list[1] are Arabic written in Hebrew characters; the rest have Hebrew star names. The rete is very uneven in thickness, with rough file marks on the reverse. The ecliptic circle is not very accurately laid out.

There are three hammered brass tympans with tangs. The first is labeled Bologna and Paris; the second is for 44° (in Arabic) and 40° (unlabeled); and the third shows only the Tropics, the equator, and the meridian on one side and is blank on the reverse.

The alidade, rule, bolt, and horse are all modern replacements.

Dr. Bernard Goldstein of the University of Pittsburgh describes this astrolabe fully in his article.[2] He has kindly summarized his comments for this catalogue:

Reverse of No. 7

"Astrolabes inscribed in Hebrew are quite rare: in addition to the Adler's, only three complete instruments, plus a rete that is now part of an Arabic instrument,[3] are known. None of the four complete instruments is signed or dated. On the other hand, many treatises on the astrolabe were written in Hebrew in the Middle Ages, and a number of star lists in Hebrew are preserved as well.[4] Still, the star names inscribed on this astrolabe do not agree exactly with any list known from a literary source.

"The medieval Hebrew astronomical tradition began in Spain in the twelfth century, and it was strongly influenced by Arabic antecedents. Abraham Bar Ḥiyya and Abraham Ibn Ezra, the two main scholars at that time whose translations and paraphrases of Arabic texts were of fundamental importance, both compiled star lists that were appropriate for use by astrolabe-makers. The list by Bar Ḥiyya depended on the star catalogue of al-Battānī (d. 929), and the list by Ibn Ezra is closely related to a list ascribed to al-Zarqāllu (11th century). The earliest known astrolabe inscribed in Hebrew, now in the British Museum, seems to date from the fourteenth century. In Levi ben Gerson's *Astronomy*

★ STAR LIST

No.	*Hebrew*	*Identification*
1	אלפארת	? Alpha Serpentis
2	נזר צפוני	Alpha Coronae Borealis
3	ארכובת התרנגלת	? Zeta Cygni
4	נשר מעופף	Alpha Aquilae
5	נסמ׳ הרמח	Alpha Bootis
6	בנות ע׳ש	Eta Ursae Majoris
7	מוש׳ הרסן	Alpha Aurigae
8	ימין הסוס	Beta Pegasi
9	לב האריה	Alpha Leonis
10	כלב קטן	Alpha Canis Minoris
11	שמאל תאומים	Gamma Orionis
12	אלדבראן	Alpha Tauri
13	נסמ׳ לא מזוין	Alpha Virginis
14	כנף הערב	Gamma Corvi
15	כוכב לפני הא[כ]וש	? Alpha Hydrae
16	כלב גדול	Alpha Canis Majoris
17	ראש הגבור	Alpha Orionis
18	רגל תאומים	Beta Orionis
19	אחרית הנהר	Theta Eridani
20	כוכב על שפ׳ המים	Delta Aquarii

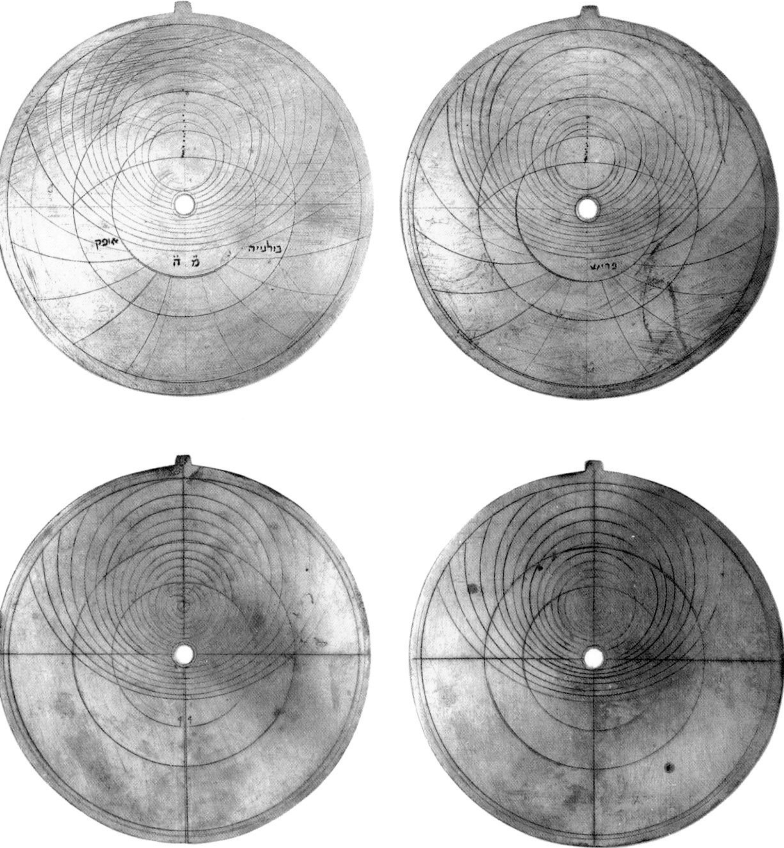

TOP: *Tympan #1 of No. 7, left for Bologna, right for Paris*
BOTTOM: *Tympan #2 of No. 7, left for 44°, right for 40°*

(*c.* 1340), the most original and comprehensive treatise on the subject written in Hebrew in the Middle Ages, there is a chapter devoted to three sources of errors in the determination of stellar altitude using an astrolabe; Levi then goes on to discuss ways to avoid these errors.[5] Of special interest in that chapter is Levi's description of the transversal scale that he invented for improving the precision of observations, made with an astrolabe, of stellar altitudes. The earliest extant transversal scale on an astrolabe is found on an instrument inscribed in Latin, dated 1483; none of the astrolabes inscribed in Hebrew has such a scale."

DATE ACCORDING TO PRECESSION — *c.* 1555.
1 ARIES = 10.25 March, from the calendar scales.
PROVENANCE — A. W. M. Mensing, Amsterdam, 1924; Max Adler, Chicago, 1930; A. P. gift, 1930.
EXHIBITIONS — *Convivencia: Jews, Muslims and Christians in Medieval Spain,* The Jewish Museum, New York, Sept. 20 - Dec. 20, 1992; The Meadow Museum, Dallas, Jan. 29 - April 11, 1993.
REFERENCES — Engelmann Catalogue (1924), 13, plate 6; Fischer *et al.* (1988), 253-92; Gibbs *et al.* (1973), 20, 31, 71, 88; Gibbs with Saliba (1984), 207, 227; Goldstein (1976), 251-60; Goldstein (1985a), 162-70; Goldstein (1985c), 185-208; Goldstein and Saliba (1983), 21, 25; Gunther (1932), 304; Mann *et al.* Catalogue (1992), 82, 88, 242; Price (1955), 363.

1. Goldstein (1976), 253.
2. *Ibid.*
3. Goldstein and Saliba (1983), 19-28; Goldstein and Chabás (1996), 317-34.
4. Goldstein (1985c), 185-208; Fischer *et al.* (1988), 253-92.
5. Goldstein (1985a), 162-70.

8 Multiple-type astrolabe

Gualterus Arsenius
Louvain
1558
"Gualterus Arscenius Louanij faciebat 1558."
on the right side of the regula
"Tentanda via est Don Luis Delacerda"
on the left side of the regula
52.2 x 39.5 x 5.0 cm.
Rete thickness — 0.6 cm.
Brass
ICA 226
M-23

Face of No. 8

This astrolabe is engraved in the Italianate script, which was promoted by Gerard Mercator (1512-1594), an engraver and cartographer who, in 1536, joined the workshop of Gemma Frisius, the uncle of Gualterus Arsenius. Most of the "y"s are written as "ij"s on this astrolabe.

The mater, which is of hammered brass, consists of a ring that is formed by a laminate of three pieces riveted together and then to the backplate. The throne is composed of elaborate, interlocking strapwork. It is dovetailed and brazed into the mater. The bail is decorated with a rosette. Further, it has a pivot and ring that appear to be original. The face of the mater carries an hour scale, labeled XII to XII twice, on the limb. Inside this scale is a circle of degrees, divided every degree, marked every five degrees, and labeled every ten degrees, 0° to 360°, starting at VI on the hour scale or Oriens on the windrose and running counter-clockwise. The cavity has a slot at the top for the tympan tangs. A small hole was drilled through the back, on the north-south line.

A *quadratum nauticum,* which is typical of Gemma's astrolabes, is engraved in the bottom of the cavity. Around the square, starting at the north (the top), are the following inscriptions, reading clockwise: "SEPTENTRIO," "Longitudo minor vel Occidentalior," "Longitudo maior siue Orientalior," "ORIENS," "MERIDIES," "OCCIDENS," "Latitudo minor vel Australior," "Latitudo maior aut Borealis."

The square is divided every two degrees, labeled by tens, 90°-0°-90°, on each side. Within the square is the windrose in Dutch, Italian, and Latin. There are 32 directional lines radiating from the center point to the edge of the square. Starting from the midpoint of "SEPTENTRIO" and moving clockwise, every other line is inscribed (◎).

The two Xalogs have a terminal flourish, similar to a "3," which must

Reverse of No. 8

represent an abbreviated ending. The flourish or tilde over the first "e" in "Poniēte maestre" indicates a missing "n," which was common practice in the Louvain school to indicate a missing letter.

Within the square and around the windrose are more winds. They are listed, clockwise, starting from the north (❖).

The square itself is divided every two degrees and marked and labeled every ten degrees starting on the north line, 0° to 90°, four times.

The rete is a typical Arsenius-type, formed by interlocking and overlapping strapwork. The ecliptic circle is divided by degrees, marked every five degrees, and labeled every ten degrees, 10°, 20°, 30°, twelve times. The zodiacal signs and names are shown, and each sign is numbered I to XII, starting with Aries, except for Cancer, which uses an Arabic four due to space constraints. The name "Vindemiatrix" is engraved on both sides of the rete with an asterisk rather than a number. The back of the ecliptic circle is divided by degrees, with the signs shown. What remains of the equator is also divided by degrees. There are 47 star points, of which 40 and Ursa maior are named (★). Most of them show astrological signs and magnitudes. Thirty-five are numbered on the back of the rete, plus one, no. 13, which has no star point.[1]

The reverse carries an Azarquiel or Gemma Frisius projection[2] on which thirteen stars have been located (☆). The Tropic of Cancer is numbered at the meridians 1 to 12 and labeled "Horae ante meridie"; the Tropic of Capricorn is numbered 12 to 1 and labeled "Horae post meridie." The ecliptic is divided and marked with appropriate symbols.

The one remaining tympan, which carries a tang, is for 37° and 41°. The 37° side is labeled "Altitudinum 37." The almucantars are drawn every degree and labeled every ten degrees. The azimuths are drawn every five

★ STAR LIST

No.	Star name	Mag.	Modern name
42	Pegasi ala		Alpha Pegasi
30	Pegasi vmbi:		Alpha Andromedae
28	Pegasi humerus	2	Eta Pegasi
27	Pegasi crus	2	Beta Pegasi
4	Crus aquarij	3	Delta Aquarii
62	Pegasi rictus	2	Epsilon Pegasi
46	Cauda capricorni	3	Gamma Capricorni
19	Cauda cijgni	2	Alpha Cygni
26	Aquila	2	Alpha Aquilae
18	Lijra	1	Alpha Lyrae
11	Caput draconis	3	Gamma Draconis
89	Ophiuchi ma: dex:	4	? Beta Ophiuchi
24	Ophiuc: cap:	2	Alpha Ophiuchi
17	Caput herculis	3	Alpha Herculis
	Scorpij cor [bent]	2	Alpha Scorpii
25	Ophiuchi sinister ma:		Delta Ophiuchi
16	Corona sept:	2	Alpha Coronae Borealis
1	Lanx boree cla:	2	Beta Librae
14	Bootis sinister hu:	3	Gamma Bootis
15	Arcturus	1	Alpha Bootis
	Ursa maior	2	Eta Ursae Majoris
	"		Zeta Ursae Majoris
	"		Epsilon Ursae Majoris
	"		Delta Ursae Majoris
	"		Gamma Ursae Majoris
	"		Beta Ursae Majoris
	"		Alpha Ursae Majoris
40	Spica virginis	2	Alpha Virginis
*	Vindemiatrix		Epsilon Virginis
	Corui ala dextra	3	Gamma Corvi
37	Cauda ♌		Beta Leonis
39	Dorsū leonis		Delta Leonis
59	Crateris fund^{9}	4	Alpha Crateris
36	Cor leonis	1	Alpha Leonis
58	Hijdre clara	2	Alpha Hydrae
	Canicula	1	Alpha Canis Minoris
56	Canis maior	1	Alpha Canis Majoris
52	Orion: dexter hūer^{9}	1	Alpha Orionis
54	Orionis sinis: pes	1	Beta Orionis
23	Hircus	1	Alpha Aurigae
32	Oculus tauri	1	Alpha Tauri
22	Meduse caput		Beta Persei
49	Nares ceti	3	Alpha Ceti
60	Venter ceti	3	Zeta Ceti
	Cauda ceti	3	Beta Ceti
31	Medi: cing: andro:		Beta Andromedae
20	Pect^{9} cassi		Alpha Cassiopeiae

The symbol "^{9}" is the Gothic shorthand for the ending "us."

☆ STAR LIST ON REVERSE

Star name	Modern name
Cor leonis	Alpha Leonis
Caput meduse	Beta Persei
Oculus ♉	Alpha Tauri
Hircus	Alpha Aurigae
Canicula	Alpha Canis Minoris
Canis maior	Alpha Canis Majoris
Dex: hūe: orioīs	Alpha Orionis
Canopus	Alpha Carinae
Spica vir:	Alpha Virginis
Arcturus	Alpha Bootis
Aquila	Alpha Aquilae
Scorpij cor	Alpha Scorpii
Lijra	Alpha Lyrae

degrees and labeled every fifteen degrees. The Tropics and the equator are labeled, as are "Horizon Rectus" and "Horizon Oblique." The crepuscular line is labeled "Linea Aurore." Also shown are the unequal hours and Great Houses, both labeled with Arabic numerals. "Oriens" and "Occidens" are shown on the left and right sides. The reverse is similarly laid out but is drawn for 41°. One of the missing tympans must have had an equal/unequal hour diagram similar to M-25's (cat. no. 10's) tympan #5 because the alidade carries the scale to effect this conversion.

❖ WINDS

*

Aparitias

*

Aquilo

Greco

*

Cecias

*

Subsolanus

*

Vulturnus

Sÿroccho

*

Euro auster

*

Auster

*

Libonotus

Lebeccio

*

Africus

*

Zephirus

*

Corus

Magistralis

*

Circius

Mater of No. 8

The regula is dovetailed to accommodate a sliding cursor with its articulated, sliding brachiolus. These last two are missing on this instrument. The regula carries the dedication and the signature, both engraved, and a scale of 360°, starting and ending at the center. This scale is for use with the ecliptic and equinoctial lines. A part iron, part brass post is set into and brazed to the regula. It carries a double groove for a U-shaped clip or horse, which is present.

The alidade is counterchanged with a beveled edge. It has fixed sights with slits and two holes, one above the other. One half is labeled, on the outer section, "Horae Ortus" with the solar symbol, ⊙. It has its scale on the beveled edge, labeled 1 to 12 and running toward the center. The inner section of this half of the alidade is labeled "Horae Occasus" with the solar symbol, ⊙, using the scale on the flat top, labeled 1 to 12 and running toward the outer end. It is used to convert from equal to unequal hours or vice versa.

The other half of the alidade has "Latitudo Meridienalis" engraved outside the equator line. This scale, on the beveled edge, is divided by degrees, marked by fives, and labeled by tens from the equator line to the end. The other section is "Latitudo

COMPASS ROSE

NOORDEN Tramõtana
Grego tramota
Grego
Griego leuante
OOSTEN Leuante
Xalog3 Leuante
Xalogue
Xalog3 medio jorno
ZUŸDEN Medio jorno
Lebeche medio jorno
Lebeche
Poniente lebeche
WESTE Poniente
Poniẽte maestre
Maestral
Maestre tramõtana

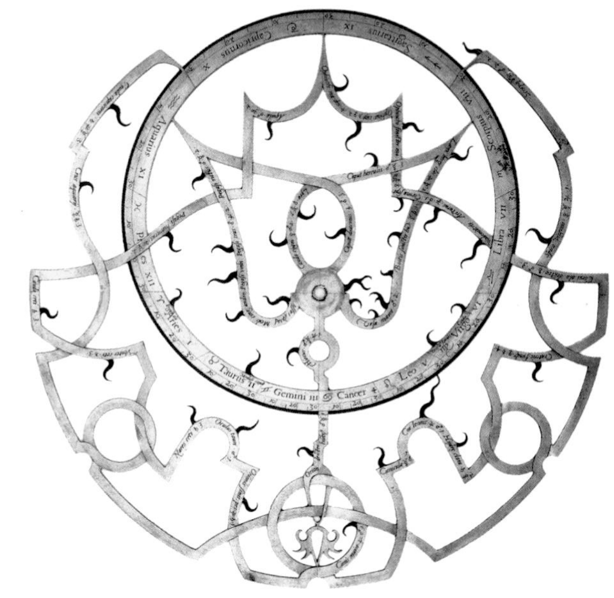

TOP: *Rete of No. 8*
BOTTOM: *Tympan of No. 8, left for 37°, right for 41°*

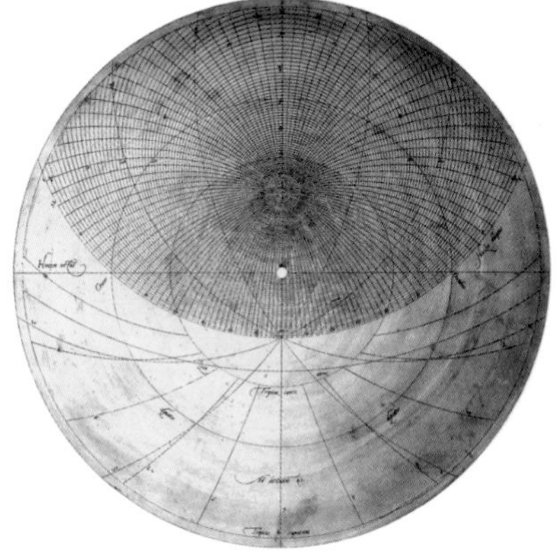

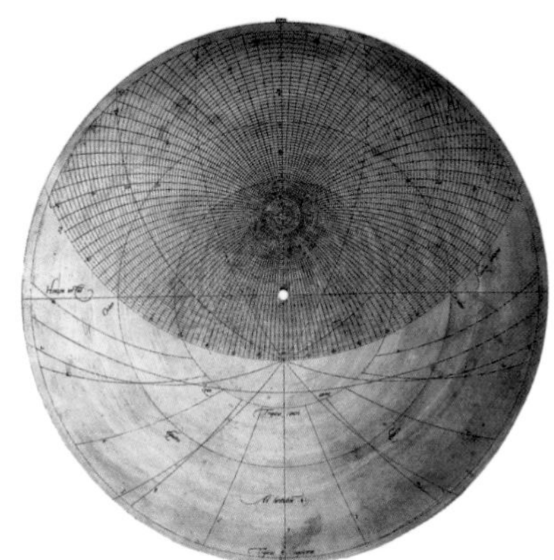

Septẽtrionalis," and its scale reads from the equator line to the center. It is divided every degree and labeled by tens.

This astrolabe was made by Gualterus Arsenius in 1558 and was inscribed to Don Luis Delacerda, probably a relative of both Gaston de la Cerda, the third Duke of Medinaceli, and Juan de la Cerda y de Silva, the fourth Duke, who was the co-governor of Flanders in 1573 with the Duke of Alba.[3]

DATES ACCORDING TO PRECESSION — *c.* 1608 on the rete and *c.* 1591 on the universal projection.
PROVENANCE — Roussel sale, Paris, 1911; Raoul Heilbronner, Paris, 1911; French government, 1914; A. W. M. Mensing, Amsterdam, 1922; Max Adler, Chicago, 1930; A. P. gift, 1930.
EXHIBITION — *Exposition Universelle et Internationale de Bruxelles,* 1935.
REFERENCES — Brussels Catalogue (1935), 4: 78, item 1815; Engelmann Catalogue (1924), 14, plates 1, 6; Fernandez Villars (1976), 7; Fox (1933), 35, 37; Gibbs *et al.* (1973), 14; Gunther (1932), 383, plate 95; Michel (1935), 8, fig. 6; Price (1955), 253, 259; Roussel Catalogue (1911), 41, lot 196; Zinner (1965), 237.

1. See appendix, p. 160, for a comparison of the numbering on the reverse of the retes of M-23 (cat. no. 8) and M-24 (cat. no. 9).
2. See p. 37.
3. Communication from F. T. Albertos, Jan. 26, 1991; Brussels Catalogue (1935), 4: 78.

9 Multiple-type astrolabe

Gualterus Arsenius
Louvain
1564
"Gualterus Nepos Gemmae F. Louvanÿ fecit 1564"
on the face, at the base of the throne
46.1 x 33.7 x 5.3 cm.
Rete thickness — 0.7 cm.
Brass
ICA 227
M-24

Face of No. 9

This astrolabe, like M-23 (cat. no. 8) and M-25 (cat. no. 10), has a classic-type face and an Azarquiel or Gemma Frisius projection on the reverse. It is also engraved in the Italianate script, but unlike M-23 the "y"s are of conventional form.

The mater consists of a cast brass ring riveted to a hammered brass backplate. The throne shows a blank shield supported on either side by two reclining figures, male and female, whose legs merge into a conventional scrolled bracket. The strap at the base of the throne, which carries the signature and date, is screwed to the mater. The face of the mater has an hour scale labeled XII to XII twice from the top. The inner scale is divided every half degree and labeled every ten degrees, starting at VI on the right side and running counter-clockwise from 0° to 360°.

The cavity in the mater has two notches, top and bottom, to hold the tangs of the tympans, which are missing. There is a *quadratum nauticum,* labeled "Quadratu Nauticum," at the top. Next, the directions appear, starting at the top and running clockwise: "MERIDIES" (south), "OCCIDENS," "SEPTĒTRIO," and "ORIENS." Then there is the square, divided every two degrees and labeled every ten degrees, 90°-0°-90°, across each side. The square itself is labeled on each side, from the top clockwise: "Longitudo minor aut Occidentalior," "Latitudo maior vel Borealior," "Longitudo maior siue Orientalior," and "Latitudo minor vel Australior."

Within the corners are engraved "Syrocho," "Lebeccio," "Magistralis," and "Greco," running clockwise from the upper left corner. There is a 32-point compass rose with a double circle engraved around it. The winds in the circle start at South (the top) and run clockwise (❖).

Within the compass rose, every other point is labeled inward, clockwise from the top (✪).

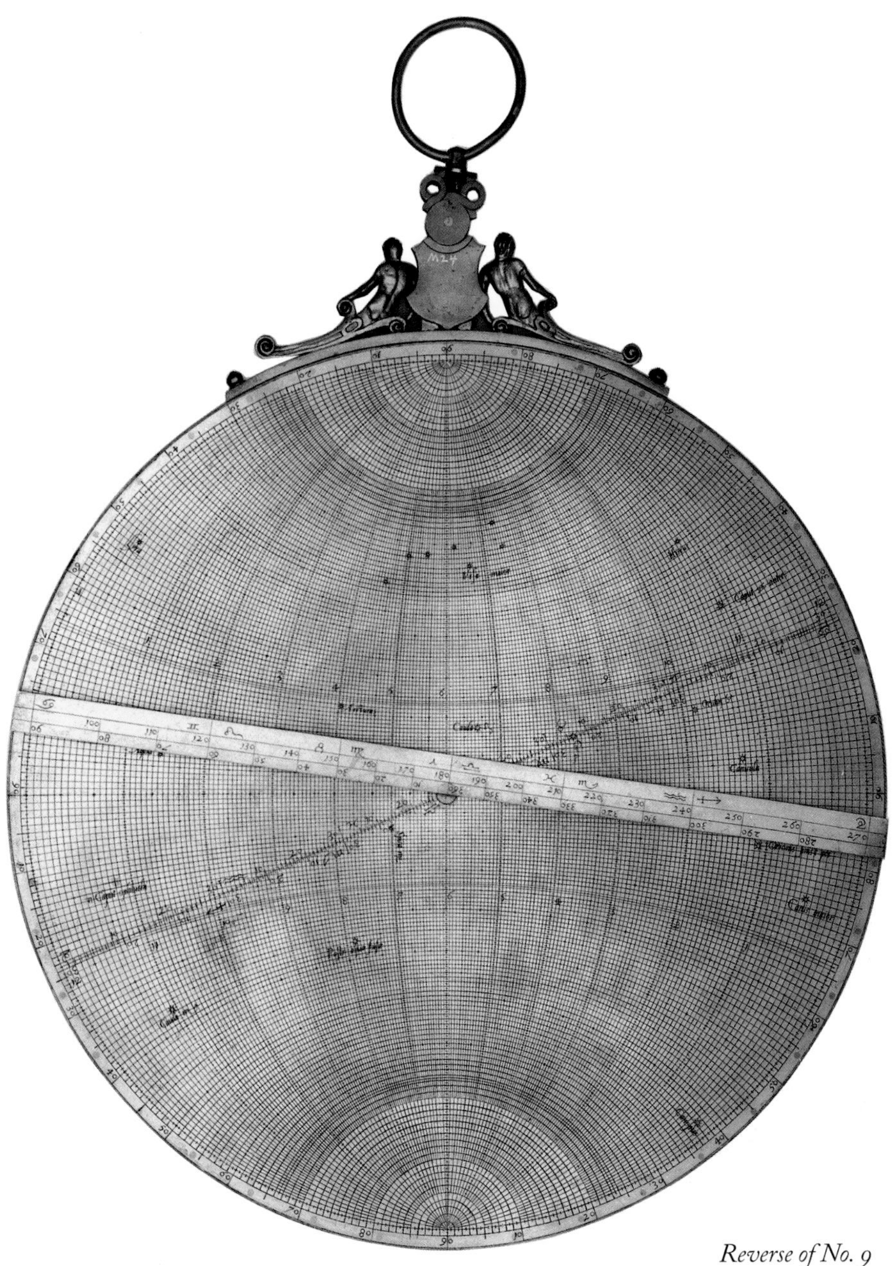

Reverse of No. 9

COMPASS ROSE

Zuÿden Medio iorno
Lebeche medio iorno
Zuÿd West Lebeche
Poniente lebeche
West Poniente
Poniente maeistre
Noort west Maeistral
Maistre tramõtana
Noordt Tramontana
Grego tramontana
Nordt oost Grego
Grego leuante
Oost Leuante
Xalogue leuante
Zuÿd oest Xalogue
Xalogue medio iorno

In engraving "Maistre tramontana," the first "n" was omitted and a tilde put above the space. The area around the center hole has been pricked to reduce its diameter.

The rete has overlapping strapwork with flame-shaped star points. There are 52 stars shown, most having their astrological signs and magnitudes (★). Ursa Major has seven star points, all unnumbered.

The back of the rete shows numbers for 45 of the star points.[1] The ecliptic circle and equator are shown as on M-23 (cat. no. 8), and the Tropic of Capricorn is laid out. Four of the seven stars in Ursa Major are marked with their magnitudes.

One of the missing tympans must have had the variant equal/unequal hour scale similar to tympan #5 of M-25 (cat. no. 10) because the alidade carries the scale to effect this conversion.

The reverse carries an Azarquiel or Gemma Frisius projection. The outer scale is divided every degree and labeled 90° to 0° four times from the top clockwise. Twenty-three stars are shown, and one position has been corrected (☆). The Tropics of Cancer and Capricorn are numbered at the meridians 1 to 12 and 12 to 1. The line

★ STAR LIST

No.	Star name	Mag.	Modern name
53	Ala pegasi	2	Alpha Pegasi
39	Pegasi vmbil:		Alpha Andromedae
40	Crus pegasi	2	Beta Pegasi
41	Pegasi humerus	2	Eta Pegasi
2	Crus ♒	3	Delta Aquarii
42	Pegasi rictus	2	Epsilon Pegasi
1	Cauda ♑	3	Gamma Capricorni
48	Cephei dex: hum:		Alpha Cephei
43	Cauda cÿgni	2	Alpha Cygni
44	Aquila	2	Alpha Aquilae
45	Lÿra	1	Alpha Lyrae
20	Caput draconis	3	Gamma Draconis
17	Ophiuc: ma: dex:	4	? Beta Ophiuchi
18	Ophiuchi caput	2	Alpha Ophiuchi
62	Ophiuchi genu dex:	3	? Eta Ophiuchi
19	Caput herculis	3	Alpha Herculis
16	Cor ♏	2	Alpha Scorpii
21	Ophiuc: sin: manus	3	Delta Ophiuchi
22	Corona Septent:	2	Alpha Coronae Borealis
23	Lanx boree clarior	2	Beta Librae
24	Bootis sinist: hum:	3	Gamma Bootis
25	Arcturus	1	Alpha Bootis
	Ursa maior	2	Eta Ursae Majoris
	"		Zeta Ursae Majoris
	"		Epsilon Ursae Majoris
	"		Delta Ursae Majoris
	"		Gamma Ursae Majoris
	"		Beta Ursae Majoris
	"		Alpha Ursae Majoris
15	Spica ♍	1	Alpha Virginis
58	Corui ala dextra	3	Gamma Corvi
33	Cauda ♌	1	Beta Leonis
34	Dorsum ♌	2	Delta Leonis
14	Crateris fundus	4	Alpha Crateris
12	Cor ♌	1	Alpha Leonis
13	Hÿdrae clara	2	Alpha Hydrae
11	Canicula	1	Alpha Canis Minoris
56	Caput ♊ anter:	2	Alpha Geminorum
10	Canis maior	1	Alpha Canis Majoris
9	Orionis dex: hum:	1	Alpha Orionis
65	Cingu: orioni media	2	Epsilon Orionis
7	Orion: sinister hum:	2	Gamma Orionis
8	Orio: sin: pes	1	Beta Orionis
35	Hircus	1	Alpha Aurigae
6	Oculus ♉	1	Alpha Tauri
51	Pers: lat^9 dex:		Alpha Persei
5	Nares ceti	3	Alpha Ceti
36	Meduse caput	2	Beta Persei
4	Venter ceti	3	Zeta Ceti
49	Andro: vmbilic^9		Beta Andromedae
3	Cauda ceti	3	Beta Ceti
37	Pect^9 cassi:		Alpha Cassiopeiae

The symbol "9" is the Gothic shorthand for the ending "us."

Rete of No. 9

of the zodiac is divided and marked with the appropriate symbols. Caput Ophiuchi was first engraved at -14° declination. Its star was then hammered from the back and filed smooth to erase the mistake. The meridian was re-engraved and the star was entered at +14° declination on the same meridian. The symbol "n" was used to mark both places.

The alidade is counterchanged with a beveled edge. One half is labeled, on the outer section, "Horae Ortus ⊙." It has its scale, labeled 1 to 10, running toward the center. The inner section of this half is labeled "Horae Occasus ⊙," and its scale runs outward from the center, labeled

WINDS

Auster
Libonothus
Africus
Zephÿrus
Corus
Circius
Aparitias
Aquilo
Coecias
Subsolanus
Vulturnus
Euro auster

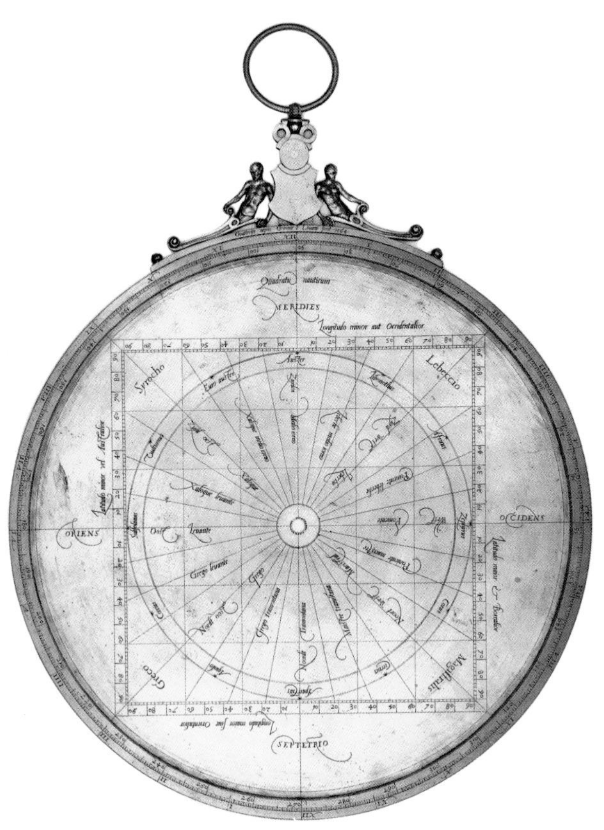

Mater of No. 9

☆ STAR LIST ON REVERSE

Star name	*Modern name*
Ursa maior	Eta Ursae Majoris
"	Zeta Ursae Majoris
"	Epsilon Ursae Majoris
"	Delta Ursae Majoris
"	Gamma Ursae Majoris
"	Beta Ursae Majoris
"	Alpha Ursae Majoris
Cauda ♌	Beta Leonis
Cor ♌	Alpha Leonis
Oculus ♉	Alpha Tauri
Canicula	Alpha Canis Minoris
Hircus	Alpha Aurigae
Caput ♊ anter:	Alpha Geminorum
Orionis sinist pes	Beta Orionis
Canis maior	Alpha Canis Majoris
Canopus	Alpha Carinae
Spica ♍	Alpha Virginis
Postr: aquae fusae	Alpha Eridani
Arcturus	Alpha Bootis
Aquila	Alpha Aquilae
Cauda ♏	Lambda Scorpii
Caput ophiuchi	Alpha Ophiuchi
Lyra	Alpha Lyrae

1 to 12. The other half of the alidade is labeled "Declinatio Septent," and its scale runs from the equatorial line to the center, labeled 10 to 70 by tens. The outer part is labeled "Declinatio Merid," and its scale runs from the equatorial line to the tip, labeled 10 to 20.

The regula is dovetailed to accommodate the missing cursor with its brachiolus. It carries a scale of 360°, starting and ending at the center, for use with the ecliptic and equinoctial lines. The ecliptic line is also shown with two symbols in each space, starting with Aries and Libra at the center. The bolt and clip arrangement is similar to M-23.

DATES ACCORDING TO PRECESSION — *c.* 1604 on the rete and *c.* 1575 on the universal projection.

PROVENANCE — A. W. M. Mensing, Amsterdam, 1924; Max Adler, Chicago, 1930; A. P. gift, 1930.

REFERENCES — Engelmann Catalogue (1924), 14, plates 1, 6; Fernandez Villars (1976), 7; Gibbs *et al.* (1973), 14; Gunther (1932), 383, plate 95; Michel (1935), 9; Price (1955), 253; Zinner (1965), 237.

1. Gerard Turner has examined around 20 astrolabe retes that are numbered on the back, and many lists agree with the numbering on M-24. We wish to thank him for his suggestion that "Corui ala dextra" was no. 58, not no. 54.

See appendix, p. 160, for a comparison of the numbering on the reverse of the retes of M-23 (cat. no. 8) and M-24 (cat. no. 9).

10 Multiple-type astrolabe

Louvain
c. 1600
Unsigned
43.2 x 31.0 x 4.0 cm.
Rete thickness — 0.8 cm.
Brass
ICA 236
M-25

Face of No. 10

The mater consists of a cast brass ring, riveted to a hammered brass backplate. It is engraved in Italianate script. The throne is composed of two dolphins or marine creatures facing each other. Their tails end in scrolls. Human faces appear on both sides of the boss between them. The bail, which has an engraved rosette as decoration, is pivoted and carries a swivel and ring. The bracket holding the throne is screwed to the mater.

The limb carries a scale that is divided every half degree, or every two minutes. The limb is labeled XII to XII twice and 0° to 90° four times, both from the top clockwise.

The cavity has a notch for the tangs of the tympans and carries a *quadratum nauticum* labeled "Quadratum Nauticū." Each side of the square is divided every two degrees and labeled by tens, 90°-0°-90°. Above the top line is inscribed "Longitudo minor siue occident" and "Longitudo maior seu orient" within faint guidelines. Along the west edge, "Longitudo minor uel Aust" and "Longitudo maior aut Boreal" also appear within faint guidelines. These last two appear to be incorrectly engraved, as "Latitudo" should have been used.

Within the square are three concentric circles and a compass rose with sixteen points (☼). These are labeled from the north, clockwise, all within faint guidelines.

The rete has the typical Louvain-type overlapping strapwork pattern with serpentine star points. There are 30 named star positions (★). The reverse of the rete shows remnants of the equator, divided by degrees.

There are five tympans; the first four are for 51° and 52°, 43° and 44°, 45° and 48°, and 39° and 41°. They all show the usual lines, with the Tropics, horizons, and equator all labeled. The Great Houses and unequal hour lines are shown.

The face of the tympan #5 carries a tablet of horizons, drawn every two

COMPASS ROSE

Tramontana
Aquilo
Grecho
Cecias
Leuante
Vlturnus
Sÿrocho
Euroauster
Mezoiorno
Lÿbonoch
Lebeccio
Africus
Ponente
Corus
Magistralis
Circius

Reverse of No. 10

degrees and labeled "Horizontale Catholicũ" twice. The "HORIZO RECTVS" is so labeled, as is "Tropicus Capricornus." The equator is labeled "Equinoctialis" twice. The meridian is labeled "Linea Meridiana." The reverse carries the scales usually found on the back of a Stöffler-type astrolabe. There are two concentric calendars, with the zodiacal one labeled with both the names and the symbols. The months are labeled in Latin, with "Janvarius," "Junius," "Julius," and "Maÿus" being the variant spellings. The shadow square is divided into 60 parts per side, marked every five and ten parts and labeled by twos, 2-12-2, twice. Above the shadow square is a variant of the equal/unequal hour scale, to be used with the scale on the alidade. The equal hour scale is marked and labeled 1 to 12 twice and "Hore ante meridiem" and "Hore post meridiem." The unequal hour scale is marked 1 to 12 twice along the upper arc. Above this, the scale is divided into twelve hours, divided every four minutes, and marked every 20 and 40 minutes. It is not labeled.

This particular type of tympan is found in many Louvain astrolabes with an Azarquiel back. The Habermel astrolabe at the National Maritime Museum in Greenwich has this type of equal/unequal hour scale on the back, as does the Coignet astrolabe at the Museum Boerhaave in Leiden,[1] and the same scale appears on a Stöffler-type tympan on the astrolabe by Thomas Gemini at the Museum of the History of Science at Oxford.[2] We can assume that this tympan should be present whenever the alidade carries the hour of sunrise and sunset conversion scale and the back of the astrolabe is an Azarquiel or Gemma Frisius projection.

The reverse of M-25 carries an Azarquiel-Gemma Frisius projection and shows eleven stars (☆). The Tropics, the equator, and the ecliptic

★ STAR LIST

Star name	*Modern name*
Pegasi umbilicus	Alpha Pegasi
Pegasi Humerus	Eta Pegasi
Pegasi crus	Beta Pegasi
Crus aquarium	Delta Aquarii
Cauda capricorni	Gamma Capricorni
Aquila	Alpha Aquilae
liera	Alpha Lyrae
Ophiuci caput [bent]	Alpha Ophiuchi
Caput Draconis	Gamma Draconis
Caput Hercules	Alpha Herculis
Cor ♏	Alpha Scorpii
Corna sept	Alpha Coronae Borealis
Bootis sinister Humerus	Gamma Bootis
Arcturus	Alpha Bootis
Principiu caude urse maioris	Epsilon Ursae Majoris
Spica ♍ [broken]	Alpha Virginis
Coruiola dextra	Gamma Corvi
Cariens fundus	Alpha Crateris
Cor leonis	Alpha Leonis
lucida Hÿdre	Alpha Hydrae
Canicula	Alpha Canis Minoris
Canis maior	Canis Majoris
Orionis sinister pes	Beta Orionis
Orionis sinister Humerus	Gamma Orionis
Hircus	Alpha Aurigae
Oculus ♉	Alpha Tauri
Ceti nares	Alpha Ceti
Venter ceti	Zeta Ceti
Pectus cascio	Alpha Cassiopeiae
Cauda ceti	Beta Ceti

Rete of No. 10

line are engraved. The Tropics are labeled 1 to 12 and 12 to 1 respectively. The ecliptic is marked every degree and with longer marks for 5°, 10°, and 15°. The zodiacal symbols are drawn.

The alidade is counterchanged and carries fixed sights. One half is labeled "Hore occasus" and "Hore ortus," with the appropriately marked and labeled scales, and is for use with the equal/unequal hour diagram on tympan #5. The other half is marked to read the declination of the sun or a star and is labeled "Latitudo Septentrionalis" and "Latitudo Me"; it is marked and labeled from the equator in both directions.

The regula is dovetailed to carry the sliding cursor and brachiolus, both now missing. It is marked and labeled by tens, 0° to 360°, from the center back to the center. The ends are decoratively cut out so the scale on the limb can be read.

M-25 closely resembles the King's College astrolabe at the Whipple Museum, ICA 235.[3] The thrones are identical, and both have the same serpentine star points. One of the tympans from ICA 235 is identical to tympan #5 of M-25. ICA 235 has Dutch wind names. There is at least one other example of a similar astrolabe, which a dealer from Plymouth, England, sent to auction several years ago. Only the mater and the ecliptic circle of the rete had survived a fire.[4]

In 1979 M-25 was tested at the Lawrence Radiation Laboratory at the University of California, Berkeley, to compare the metal to that of the "Plate of Brass."[5] Three samples were taken from the inside edge of the cast ring and three from the tympan for 51° and 52°. The latter already had a small hole below the tang, and the samples were removed at the other cardinal points. Both sets of samples were analyzed by neutron activation analysis (NAA), X-ray fluorescence (XRF), and emission spectroscopy (ES). Drs. Helen

☆ STAR LIST ON REVERSE

Star name	*Modern name*
Cor ♌	Alpha Leonis
Oculus ♉	Alpha Tauri
Canis minor	Alpha Canis Minoris
Hircus	Alpha Aurigae
Orionis si: pes	Beta Orionis
Canis maior	Alpha Canis Majoris
Canopus	Alpha Carinae
Postremasu: aque	Alpha Eridani
Arcturus	Alpha Bootis
aquilo	Alpha Aquilae
liera	Alpha Lyrae

Tympan #5 of No. 10
LEFT: *Tablet of horizons*
RIGHT: *Stöffler-type scales*

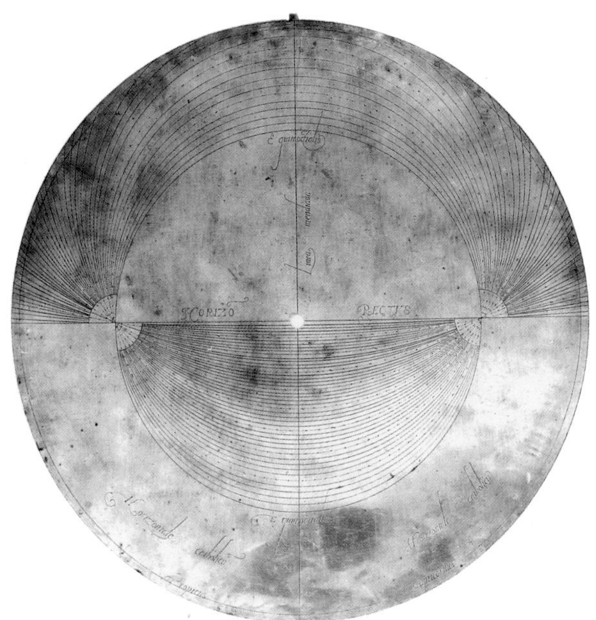

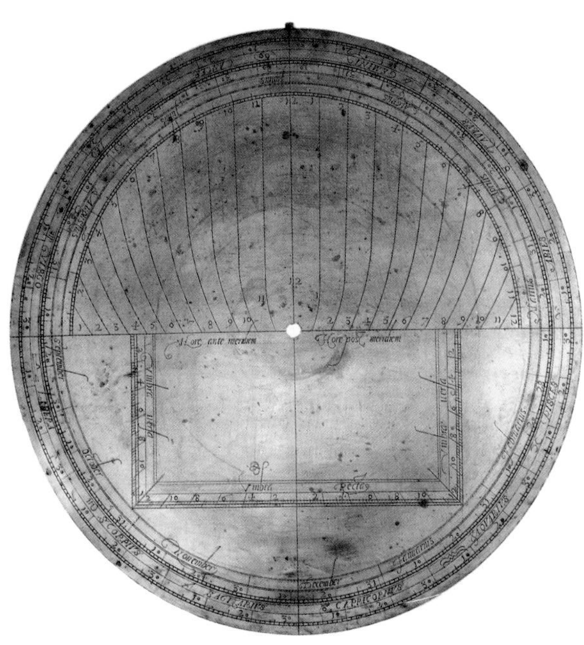

Michel and Frank Asaro were instrumental in the testing, and Dr. Arthur Norberg was the facilitator. We are deeply appreciative of their interest and labors. The composition tables are reprinted here with permission of the authors (❄).[6]

DATES ACCORDING TO PRECESSION — *c.* 1594 on the rete and *c.* 1618 on the universal projection.
I ARIES = 10.5 March, from the calendar scales.
PROVENANCE — A. W. M. Mensing, Amsterdam, 1924; Max Adler, Chicago, 1930; A. P. gift, 1930.

EXHIBITIONS — Abrams Planetarium, East Lansing, Mich., 1963-65; *The Triumph of Humanism, 1425-1625,* California Palace of the Legion of Honor, San Francisco, Oct. 27, 1977 - Jan. 8, 1978; *The Splendors of Dresden,* Fine Arts Museums of San Francisco, Feb. 8 - May 26, 1979; *Sir Francis Drake,* Bancroft Library, University of California, Berkeley, June 14 - Oct. 6, 1979.
REFERENCES — Engelmann Catalogue (1924), 14, plates 1, 6; Gibbs *et al.* (1973), 14; Gunther (1932), 394, plate 98; Gunther (1937a), plate between 186 and 187; Hart *et al.* (1979), 5, 14-17; Michel (1947), 162; Price (1955), 255; memorial volume for Dr. Isodore Perlman, University of California Press, Berkeley (pending).

1. National Maritime Museum Astrolabes (1976); Gent (1994).
2. Gunther (1932), 738-39.
3. *Ibid.,* 393-94, plate 98; Bryden Catalogue (1988), entry 345; Gunther (1937a), 186, plate between 186 and 187.
4. Photographs on file in History of Astronomy Dept., Adler Planetarium.
5. The "Plate of Brass" was found in northern California and thought, by some people, to have been left by Francis Drake.
6. Hart *et al.* (1979), 5, 14-17.

TABLE 1 COMPOSITION OF ASTROLABE

		ASTR-1 (Horizon Plate)			ASTR-2 (Tympan 51°)		
Cu[1]		68.5	±	1.5	68.4	±	1.5
Zn		30.6	±	1.2	30.9	±	1.1
Ag[3]	ppm	278.0	±	12.0	260.0	±	11.0
As	ppm	470.0	±	43.0	392.0	±	36.0
Sb	ppm	304.0	±	5.0	67.0	±	1.0
In[3]	ppm	<0.5			1.3	±	0.3
Au	ppm	1.5	±	0.3	1.3	±	0.3
Sn		0.53	±	0.10	1.0	±	0.1
Ni	ppm	2598.0	±	33.0	2413.0	±	22.0
Fe[4]		0.08	±	0.005	0.140	±	0.005
Cd	ppm	<171.0			<179.0		
Pb[2]		1.2	±	0.1	1.4	±	0.1
Mg[4]	ppm	7.0	±	2.0	5.0	±	2.0

1. Composition in % unless indicated in ppm.
2. XRF measurements done by Lily Goda and Robert Giauque.
3. 20% uncertainty due to flux monitor calibration.
4. Emission spectroscopy done by George V. Shalimoff.

TABLE 2 TRACE ELEMENT COMPOSITION[1] OF BRASSES

Sample	Date	As	Ni	Sb	Ag	Au (ppm)	Sn	Pb	Fe	Average ratio (To Plate of Brass)
Astrolabe:										
Mater	1570	0.047	0.26	0.0304	0.028	1.5	0.53	1.2	0.08	27
Tympan 51°	1570	0.039	0.24	0.0068	0.026	1.3	1.00	1.4	0.14	33
Rhineland	18th c.	0.036	0.09	0.033	0.020	2.0	1.40	2.8	1.0	48
France/Spain	14th c.	0.22	0.19	0.76	0.11	5.6	1.07	6.0	<1.2	167
Italy	17th c.	0.33	0.63	0.71	0.23	67.0	1.45	2.7	0.5	217
Italy	17th c.	0.41	0.33	0.83	0.06	18.0	1.41	3.1	0.3	198
Italy	17th c.	0.41	0.42	0.55	0.12	35.0	2.5	3.5	0.5	193
Italy	17th c.	0.24	0.34	0.43	0.21	72.0	2.25	4.3	0.5	186
Plate of Brass	19-20th c.	<.0055[2]	<0.01	0.00083	0.005	0.21	0.006	0.10	0.027	1

1. Abundances are in percent except for those of gold, which are in parts per million (ppm).
2. One half of upper limit was used in calculation.

TABLE 3 SUMMARY OF AVERAGE RATIOS OF OLD BRASSES

Geographic area	*Date (century* AD*)*	*No. of artifacts*	*Mean*	RMSD[1]
Belgium, Rhineland	16-18	3	36	11
Italy	16-17	4	198	13
France, Spain?	14	1	167	
Plate of Brass	(19-20)	1	1	

1. Root-mean square deviation.

11 Classic-type astrolabe

Bernardinus Zabeus
Padua
1559
"OPVS BERNARDINI ZABEI IN PADOVA"
on the left edge
"ANO DOMINI 1559 MENSI APRILIS"
on the right edge
17.7 x 12.5 x 3.8 cm.
Rete thickness — 1.1 cm.
Brass
ICA 177
M-21

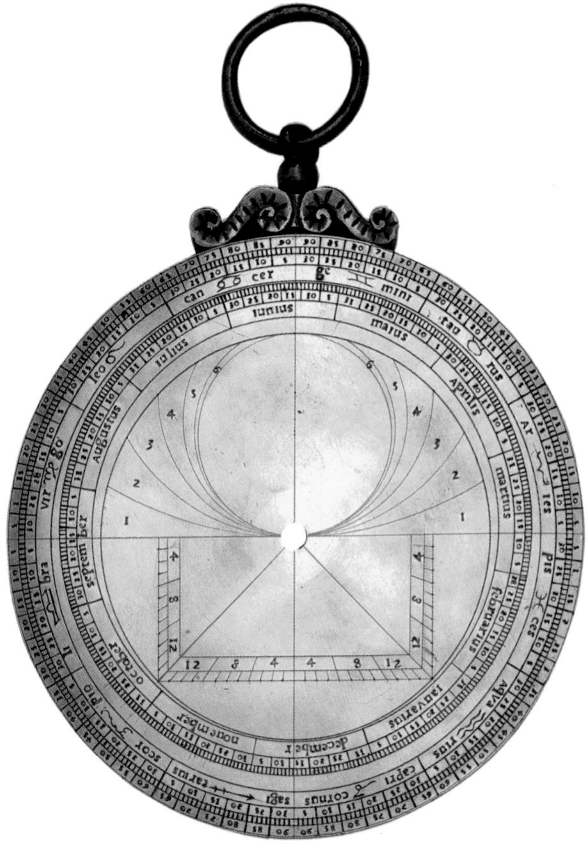

LEFT: *Face of No. 11*
RIGHT: *Reverse of No. 11*

In this astrolabe the mater is cast brass, and the cavity for the tympans has been turned out on a lathe. The throne is a pair of scrolls, decorated by short lines. It is attached to the mater by brazing or soldering. The ring and swivel-bail are present.

The face of the mater has an hour scale on the limb, divided and labeled 1 to 12 twice and also divided by degrees, which are marked and labeled every five degrees within each hour. There is a slot to accommodate the tangs of the tympans.

The back of the mater shows the conventional double calendars, which are eccentric. The month names are in Latin on the civil calendar. Both signs and full names of the zodiac are shown. Within the calendar scales there is a double unequal hour diagram, labeled 1 to 6 twice. Below this is a shadow square, divided into twelve parts and labeled every fourth part.

The rete has 32 star points (★). The reverse is blank. Regulus is indicated by a small pointer in the outer edge of the ecliptic circle.

There are five tanged tympans, four of which are labeled "G" (Gradus) 27° and 30°, 33° and 36°, 45° and 48°, and 51° and 54°, respectively. Tympan #5 has multiple horizons on one side and on the other, in a different hand,

Signature on No. 11

★ STAR LIST

Star name	*Modern name*
caput anDromaDa	Alpha Andromedae
crus aquarius	Delta Aquarii
cefeu umi	Alpha Cephei
cauDaga	Alpha Cygni
dalPhi	Alpha Delphini
vvltuvo	Alpha Aquilae
galina	Beta Cygni
draco	Gamma Draconis
v. ca	Alpha Lyrae
serPentarius	? Serpentarii
herculis ala	? Herculis
scorPio	Alpha Scorpii
coro	Alpha Coronae Borealis
Artur boet	Alpha Bootis
vrsa ma [5 star points]	Ursae Majoris
spica	Alpha Virginis
corvvs	? Corvi
idra	Alpha Hydrae
cancer	? Cancri
canis mi	Alpha Canis Minoris
canis ma	Alpha Canis Majoris
dester hume	Alpha Orionis
[unnamed]	[Gamma Orionis]
orionis	Beta Orionis
Alaior	Alpha Aurigae
taurus oculi	Alpha Tauri
[unnamed]	[? Orionis]
persei	Alpha Persei
meDuse	Beta Persei
anDro	Beta Andromedae
venter ceti	Zeta Ceti
[unnamed]	[? Ceti]

a much reworked diagram for 45°, with many false starts and erasures. On the first four tympans the almucantars are drawn and labeled every three degrees, the azimuths every ten degrees. Besides the usual lines, the crepuscular line and the Great Houses are shown.

The alidade is unmarked, and the rule is double, counterchanged, and unmarked. The astrolabe is held together by a slotted bolt and a horse.

DATE ACCORDING TO PRECESSION — *c.* 1555.
I ARIES = 10 March, from the calendar scales.
PROVENANCE — A. W. M. Mensing, Amsterdam, 1924; Max Adler, Chicago, 1930; A. P. gift, 1930.
REFERENCES — Engelmann Catalogue (1924), 13, plates 1, 6; Fox (1933), 35; Gibbs *et al.* (1973), 14; Gunther (1932), 329-30; Michel (1947), 176; Price (1955), 253.

12 Classic-type astrolabe

Laurentius Schreckenfuchs
Memmingen, Germany
1567
"M. Laurentius Schreckenfuchis, / Menningensis faciebat Anno / MDLXVII Q. / N. Q. N."
on the back, below the shadow square
14.5 x 12.7 x 0.11 cm.
Wood, paper, and vellum
W-109

❖ WINDS

Winds	*Next to*
Septend	XII
Boreas	II
Africus	V
Subsolanus	VI
Eurus	VII
Chorus	VII-VIII
Circius	IX

LEFT: *Mater of No. 12 with tympan #1 (for 48°)*
RIGHT: *Reverse of No. 12*

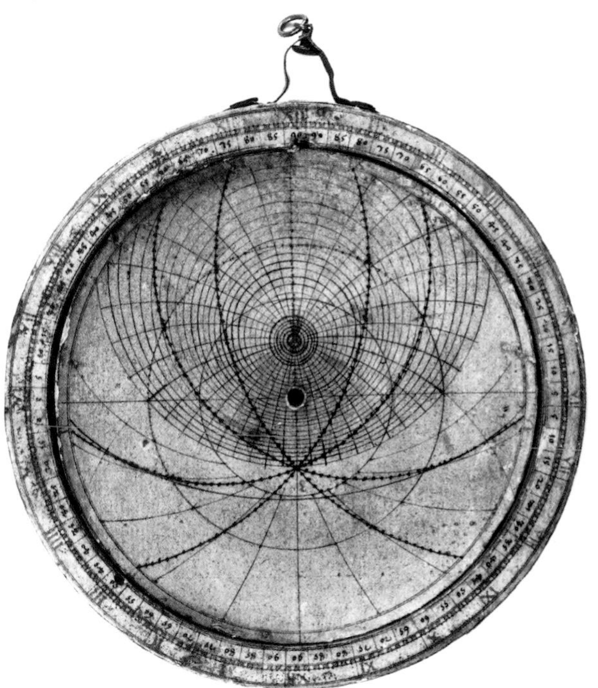

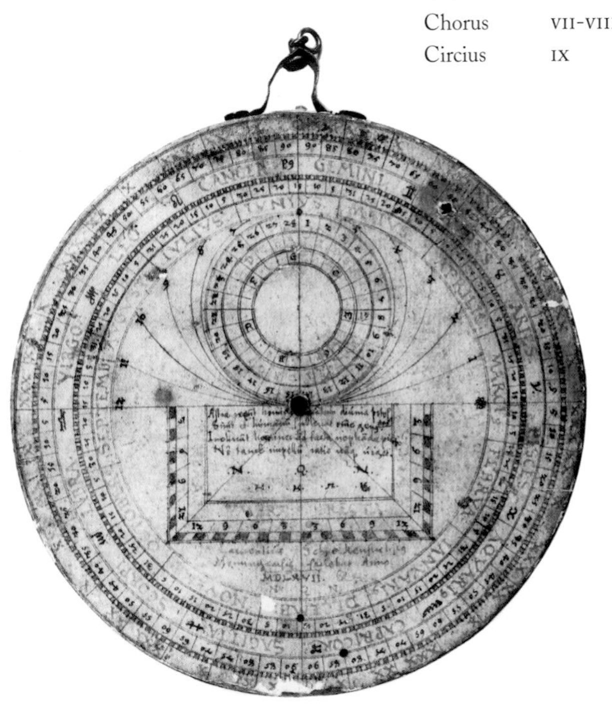

■ MONTH NAMES

IANVARI9	IVLIVS
FEBRVA	AVGVST9
MARCI9	SEPTEMB
APRILIS	OCTOBER
MAIVS	NOVEMB:
IVNIVS	DECEMB:

The symbol "9" is the Gothic shorthand for the ending "us."

The mater is a wooden ring with a paper overlay in manuscript. The backplate is also of wood with a paper overlay in manuscript. The throne is a narrow brass strip, raised in center to hold the bail; it is nailed to the mater. The ring is present.

The limb on the face carries an hour scale divided and labeled XII to XII twice, with a scale of degrees, alternately hatched, and marked and labeled every five degrees, 90°-0°-90°, twice from the top.

On the edge are the winds in Italian (❖). The other winds are illegible. The cavity is unmarked. The cardinal winds are in red.

The limb on the reverse carries two concentric calendars. From the outside in: a scale is divided by degrees, alternately hatched, labeled X-XX-XXX twelve times on the outside and marked and labeled by five degrees, 90°-0°-90°, twice from the top on the inside. Then come the zodiacal signs and names. The civil calendar is hatched as above, marked and labeled by tens, 10, 20, and 28, 30, or 31 as the case may be. The month names are in Latin (■).

Above the median line there is a double unequal hour scale, labeled 1 to 12 counter-clockwise. Within this scale are three concentric circles,

Astra _____[?] homin__ [quantum?] diurna potestas
Sunt et humanum sublevat omne genus.
Inclinant homines ad fasta nephanda piaque
Non tamen impellunt; ratio nmquam viget.[?]
N. Q. N.
H[?] K. r. K.

The stars __ [are?] __ a daily power and it aids the entire human race. They [the stars] dispose men towards [observing events pertaining to] calendar days, both profane and sacred but they do not compel them; reason never thrives.

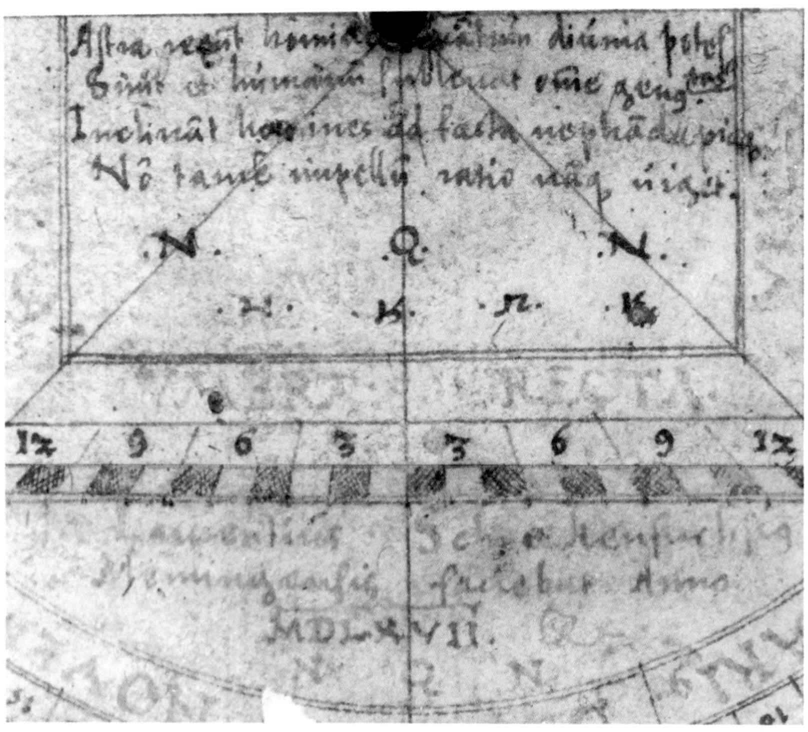

Inscriptions on No. 12

each divided into 28 parts. The outside one is labeled I to 28 in black. The middle one, in red, is mostly obliterated, but it must have been the Golden Letters, A to F. The inner one must be a leap year table as the letters are widely spaced, with three spaces between the letters. They appear to be A to G.

The shadow square is divided into twelve parts, divided by ones and labeled by threes. "VMBRA VERSA" and "VMBRA RECTA" are shown in red. Within the shadow square is a six-line inscription. Dr. Nancy Kassell of Brookline, Massachusetts, has kindly supplied the transcription and translation (●). We greatly appreciate her efforts. The signature is below the square.

The rete, alidade, rule, bolt, and horse are missing.

There are two vellum tympans with red Arabic numerals. One is for 48° and is blank on the reverse; the other is for 51° and is also blank on the reverse. The almucantars are drawn every three degrees. The unequal hour lines and Great Houses are shown. The latter are labeled I to XII and are shown by dotted lines. The unequal hour lines are labeled I to 12 clockwise on the 51° tympan.

Laurentius Schreckenfuchs may have been the son of Erasmus Oswald Schreckenfuchs (1511-1579), who wrote on astrolabes and quadrants.[1]

I ARIES = II March, from the calendar scales.
PROVENANCE — Alain Brieux, Paris, before 1983; R. S. Webster, Chicago, 1983; A. P. gift, 1983.
REFERENCE — Brieux Catalogue (1983), 29.

1. Zinner (1965), 530.

13 Surveyor's astrolabe

Germany
Before 1582
Unsigned
18.7 x 13.7 x 2.7 cm.
Rete thickness — 0.5 cm.
Brass
M-45

LEFT: *Face of No. 13*
RIGHT: *Reverse of No. 13*

This instrument has hammered brass plates soldered to each side of a cast ring, forming a drumlike mater. The throne, in the shape of a double ogee curve, is soldered to the ring. A shackle, pivot, and ring are present.

The reverse takes the form of the back of a classic astrolabe. The limb carries a circle of 360 degrees that is divided every degree, marked every five and ten degrees, and labeled 90°-0°-90° twice from the top. The same circle of degrees also serves as the zodiacal calendar, which is marked every 30 degrees and labeled with the names in Latin; "Sagitar," "Capric," and "Aquari'" are the variant spellings. Inside the zodiac is an eccentric civil calendar, misdivided into 30-day months and labeled in Latin: "IAN," "FEB," "MAR," "APR," etc.

Within the civil calendar there is a concentric circle that encloses a shadow square and an unequal hour diagram. The shadow square is divided into twelve parts, marked and labeled every fourth part, 4-8-12, 12-8-4, twice. Above the square is an unequal hour diagram, labeled 1-6-1.

The fives are Gothic in form, and the letters "S" and "N" are stamped backward.

The face has a scale on the limb of 360 degrees, divided every degree and marked every five and ten degrees but not labeled. Below the center line is a shadow square, divided into 24 parts, marked and labeled every fourth part, 4 to 24, 24 to 4, twice. The same stamps are used in marking both sides of the instrument.

Both the alidade on the reverse and the single rule on the face are original and are decorated with a delicate feather design, extending from the center.

1 ARIES = 9.5 March.
PROVENANCE — A. W. M. Mensing, Amsterdam, 1924; Max Adler, Chicago, 1930; A. P. gift, 1930.
REFERENCE — Engelmann Catalogue (1924), 17.

14 Classic-type astrolabe

Johannes Bos
Italy
1597
"IOANNES BOS. I. DIE. 24. MARTII. 1597" *on the rete*
16.0 x 12.5 x 2.2 cm.
Rete thickness — 0.5 cm.
Brass
ICA 185
M-33A

★ STAR LIST

Star name	*Mag.*	*Modern name*
AVST	3	Delta Aquarii
AQVILA V·	2	Alpha Aquilae
C· CAPR·	3	Gamma Capricorni
C· CIG·	2	Alpha Cygni
SP· LIRAE FIDICVLA		Alpha Lyrae
C· DR·	3	Gamma Draconis
C· HER·	3	Alpha Herculis
COR ♏	2	Alpha Scorpii
L· CO·	2	Alpha Coronae Borealis
ART		Alpha Bootis
SPICA· VIRG·	1	Alpha Virginis
P· C· VR· M	2	Epsilon Ursae Majoris
ALA· D· C·	3	Gamma Corvi
EL· C· V· M	2	Delta Ursae Majoris
R· CORVI	3	Alpha Corvi
IL· VR MA	2	Alpha Ursae Majoris
COR LEONIS	1	Alpha Leonis
LVCIDA IDRAE	2	Alpha Hydrae
CANIS MINOR PROCIOH	1	Alpha Canis Minoris
SIRIVS	1	Alpha Canis Majoris
D· HVM· OR	1	Alpha Orionis
SIN· PES· ORIONIS RIGEL	1	Beta Orionis
HIRCVS	1	Alpha Aurigae
ALDEB· OC· TAVRI	1	Alpha Tauri
MED· C· OR	2	Epsilon Orionis
CA AL	2	Beta Persei
P· CAS·	3	Alpha Cassiopeiae
VENTER CETI	3	Zeta Ceti

Face of No. 14

This astrolabe is one of at least five instruments signed and dated "Ioannes Bos. I. die. 24. Martii. 1597." There is obviously "something rotten in the state of Denmark." Bos, whose father was Jacob Bos, a cartographer in Antwerp,[1] was a known maker of astrolabes and calculating devices[2] who worked in Antwerp and Rome between 1591 and 1623. It is probable that Johannes Bos made an astrolabe signed and dated this way that was later copied many times over. This seems to us to be the most likely scenario as M-33A is larger than the others and is made of hammered brass, of uneven thickness. The other examples that we have seen seem to be of rolled brass. They may be found at Harvard,[3] the Museum of the History of Science at Oxford,[4] the Whipple Museum of the History of Science at Cambridge,[5] and the Museum Boerhaave in Leiden.[6]

The mater is formed of a ring with the backplate soldered to it. The throne is a delicate, open scroll that shows an early repair. The swivel and ring seem to be original. The face of the mater has an hour scale, labeled XII to XII twice and divided into degrees. These are alternately hatched and labeled 90°-0°-90° twice from the top. The cavity is unmarked but has a slot to position the tangs of the tympans.

The reverse of the mater carries two eccentric calendar scales. The signs of the zodiac are engraved. The initials of the months are in Italian, with July being shown as "L" instead of "G." There is a sunburst above the shadow square and three wind faces, on the sides and below the square.

The shadow square is divided into twelve parts alternately hatched, marked, and labeled by threes, 12-3-3-12, on all three sides. "VMBRA VERSA" is inscribed on both sides and "VMBRA RECTA" across the bottom.

The rete is a double cordiform-type surrounded by sinuous strapwork. There are 28 named star points, each showing astrological signs and magnitude numbers (★). The signature and date appear in the rete between Sagittarius and Capricorn. The ecliptic circle is divided every two degrees and is hatched, marked, and labeled by tens and thirties, with the zodiacal signs shown. The reverse is plain.

Only one tympan remains. It is for 43° and 44° and has a tang at the top. The 43° side has almucantars drawn every five degrees, with the first labeled "5°." Only the south and east-west azimuths are shown. The tympan carries the usual lines plus the Great Houses (dotted lines) and the unequal hours. The 44° side has its almucantars drawn every three degrees, with the one above the horizon labeled "3" at each end. The azimuths are drawn for every ten degrees. The Great Houses are indicated by dotted lines. The unequal hours are shown below the horizon. A carefully made, tanged filler occupied the balance of the cavity.

The alidade is counterchanged and unmarked, with folding sights. The rule is double and counterchanged. It is marked to help determine the declination of the stars. The bolt and nut are original.

There is another group of Bos astrolabes showing different dates, including ones at the National Maritime Museum in Greenwich[7] and in the former Van Alfen Collection.[8] There are, further, several other astrolabes with different signatures but having retes similar to the Bos. These include the examples at the Los Angeles County Museum[9] and at the Adler Planetarium (see A-III, cat. no. 29). Dr. Maria Rooseboom mentions four Bos astrolabes, those at the Adler, Oxford, Greenwich, and Leiden.[10]

DATE ACCORDING TO PRECESSION — *c.* 1601.
I ARIES = 21 March, from the calendar scales.
PROVENANCE — A. W. M. Mensing, Amsterdam, 1924; Max Adler, Chicago, 1930; A. P. gift, 1930.
REFERENCES — Bryden Catalogue (1988), no. 386; Engelmann Catalogue (1924), 15, plate 2; Fox (1932), 549, fig. 44; Fox (1933), 35; Gent (1994), 41, 44; Gibbs *et al.* (1973), 14; Gunther (1932), 335; Michel (1947), 164; Price (1955), 255; Price (1956), 382-83, pictures on 384-85; Rooseboom (1950), 36; Zinner (1965), 254.

1. Zinner (1965), 254.
2. Sotheby's (1952).
3. Webster notes.
4. Price (1955), 255.
5. Bryden Catalogue (1988), no. 386.
6. Zinner (1965), 254; Price (1955), 255; Gent (1994), 41, 44.
7. Price (1955), 255.
8. Milo *et al.* Catalogue (1955), III, no. 343.
9. Webster notes.
10. Rooseboom (1950), 36.

Reverse of No. 14

15 Classic-type astrolabe

Ludovicus Martinot
Sens, France
1598
"Ex mente Christophori Lauren Agendicensis • J • C • /
Ludouicus Martinot Horologopeus fecit Agendicy / 1598"
under the shadow square
33.3 x 26.4 x 3.15 cm.
Rete thickness — 0.35 cm.
Brass
ICA 122
M-31

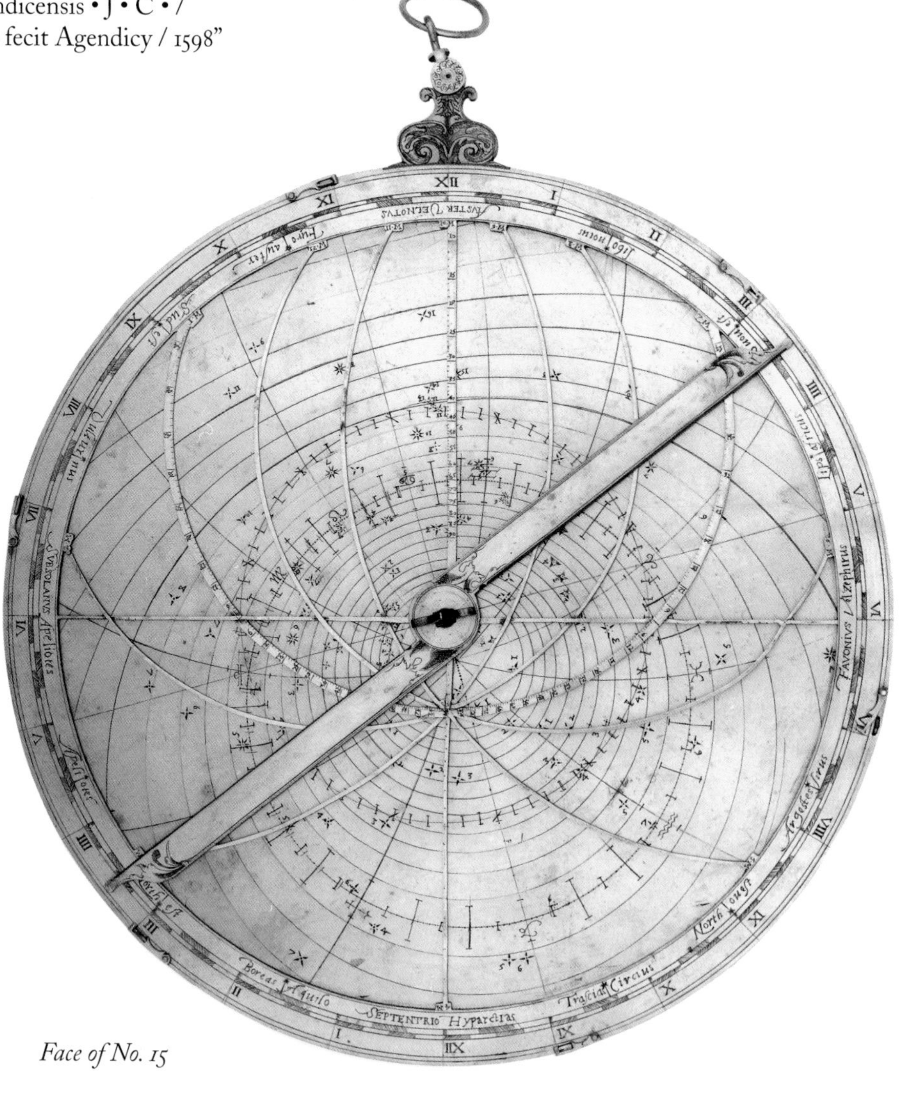

Face of No. 15

The mater consists of two parts that are latched together with hooks and posts. The face is a plate cut out to show only the limb, the prime meridian, the right horizon, the arcs of the oblique horizon, and the Great Houses.

The limb carries the hour scale, XII to XII, twice; each hour is divided into four parts and, on a secondary scale, into three parts. There is a windrose around the inner edge of the limb; four directions and twelve winds are engraved, starting at the top and running counter-clockwise (❖).

The Great Houses are numbered 1M to 12M starting at the east end of the oblique horizon and running counter-clockwise. The oblique horizon is itself divided by degrees and labeled every six degrees, 60°-0°-90°, from the rim through the right horizon to the prime meridian, with a mirror image on the other side. The prime meridian is divided by degrees and labeled every five degrees from 0° at the rim to 90° at the zenith.

The back of the mater is an annular ring carrying the throne and has five of the original posts for latching. The throne is a double scroll with a swivel-bail and ring. The limb carries two concentric calendars. The signs of the zodiac are shown; the month names are in Latin, with "Janvarius," "Junius," and "Julius," the same spellings found on M-27 (cat. no. 2), another French astrolabe.

A rotatable disc is held between the front plate and the annular ring. It carries the star chart on the face, showing more than 104 stars, 100 of which we have identified using Bayer (1603) and Blaeu's 1628 planisphere in Gunther (1932) (★). The stars within each zodiacal sector are numbered in sequence starting from the pole of the ecliptic. The numbered stars and star groups within the sectors range from six to sixteen, with the total being 104. In Taurus there are two

STAR LIST

Alpha Cephei
Delta Cephei
Alpha Ursae Minoris
Zeta Ursae Minoris
Alpha Cygni
Beta Cygni
Beta Draconis
Alpha Lyrae
Gamma Ursae Minoris
Alpha Ursae Majoris
Beta Ursae Majoris
Gamma Ursae Majoris
Delta Ursae Majoris
Epsilon Ursae Majoris
Zeta Ursae Majoris
Eta Ursae Majoris
Kappa Ursae Majoris
Theta Ursae Majoris
Alpha Aurigae
Beta Aurigae
Gamma Aurigae
Alpha Persei
Beta Persei
Alpha Trianguli
Beta Cassiopeiae
Alpha Coronae Borealis
Alpha Bootis
Gamma Bootis
Alpha Herculis
Alpha Serpentarii
Beta Ophiuchi
Delta Ophiuchi
? Ophiuchi
Alpha Scorpii
Beta Scorpii
Lambda Scorpii
Alpha Geminorum
Beta Geminorum
Alpha Andromedae
Beta Andromedae
Delta Andromedae
Gamma Andromedae
Alpha Pisces
Alpha Pegasi
Beta Pegasi
Pleiades
Alpha Tauri
Gamma Tauri
Alpha Delphini
Alpha Aquilae
Theta Aquilae
Alpha Sagittae
Alpha Equi Minoris
Beta Equi Minoris
Alpha Capricorni
Beta Capricorni
Gamma Capricorni
Delta Aquarii
Alpha Canis Minoris
Beta Canis Minoris
Alpha Canis Majoris
Alpha Orionis
Beta Orionis
Delta Orionis
Epsilon Orionis
Zeta Orionis
Lambda Orionis
Theta Orionis
Gamma Orionis
Alpha Leporis
Alpha Carinae
Alpha Crateris
Theta Crateris
Alpha Corvi
Beta Corvi
Alpha Ceti
Beta Ceti
Zeta Ceti
Epsilon Ceti
Alpha Hydrae
Gamma Eridani
Alpha Virginis
Epsilon Virginis
Zeta Virginis
Alpha Leonis
Beta Leonis
Delta Leonis
Gamma Leonis
Alpha Cancri
Gamma Cancri
Alpha Arietis
Alpha Librae
Beta Librae
Alpha Centauri
Beta Centauri
Alpha Piscis Australis
Alpha Sagittarii
Beta Sagittarii
Alpha Arae
Beta Arae

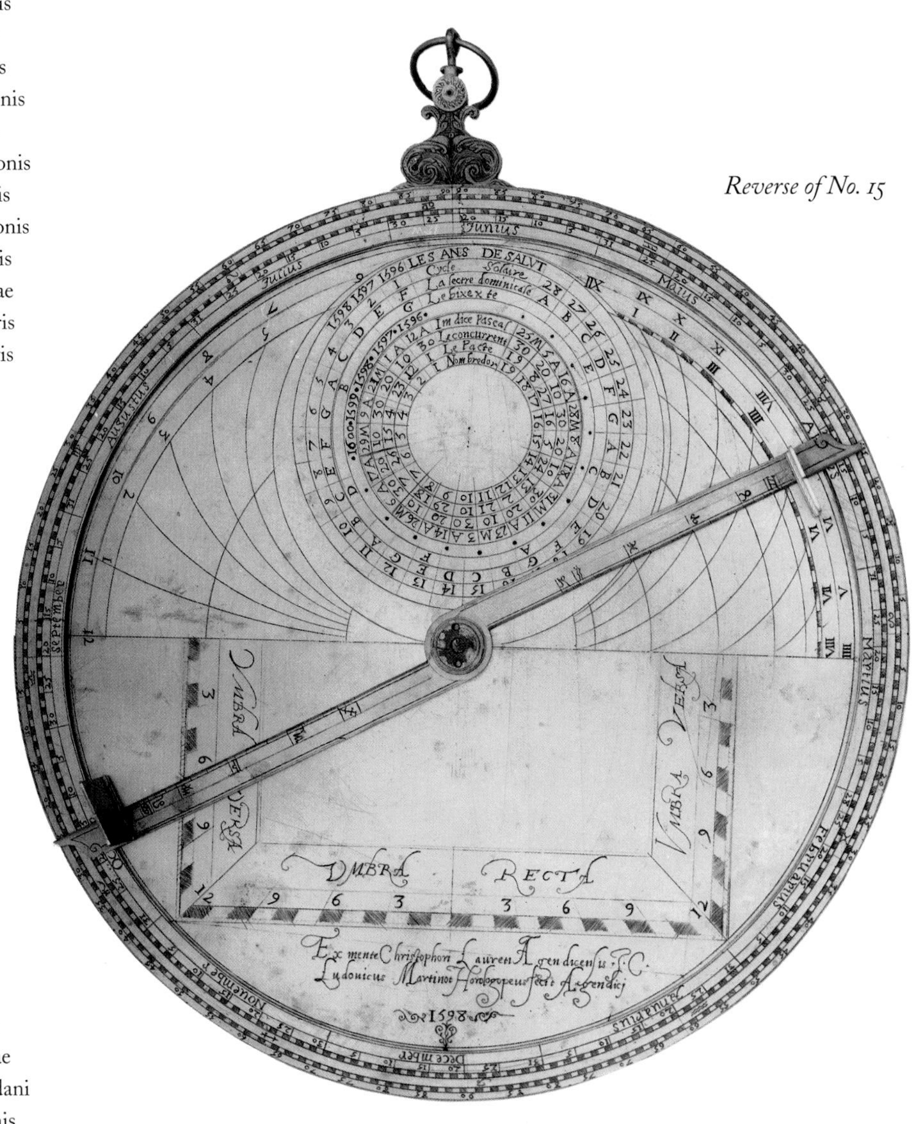

Reverse of No. 15

stars, each numbered "I." In several cases a number will represent a group of stars. The magnitudes are indicated by the form and number of rays for each star symbol. The ecliptic circle, which is divided by degrees and marked every five and ten degrees, has the signs of the zodiac engraved on it. A series of eccentric circles shows the angular distance from the pole of the ecliptic. The lines radiating from this pole mark the zodiac. The center hole marks the North Celestial Pole. The celestial equator is shown, divided every five degrees and marked every 30 degrees with an "x."

The back of the rotatable disc shows the signature, the equal/unequal hour scale, various tables, and the shadow square, which is divided into twelve parts, alternately hatched. It is marked and labeled by threes, 3-6-9-12-9-6-3, twice. It is also labeled "VMBRA VERSA" at each end and "VMBRA RECTA" across the front. The inscription and date appear below this.

The unequal hours are shown on the left in the upper half and labeled I to 6 and 6 to 12 back from the horizon. The equal hours are shown on the right, each divided into four parts, alternately hatched and labeled

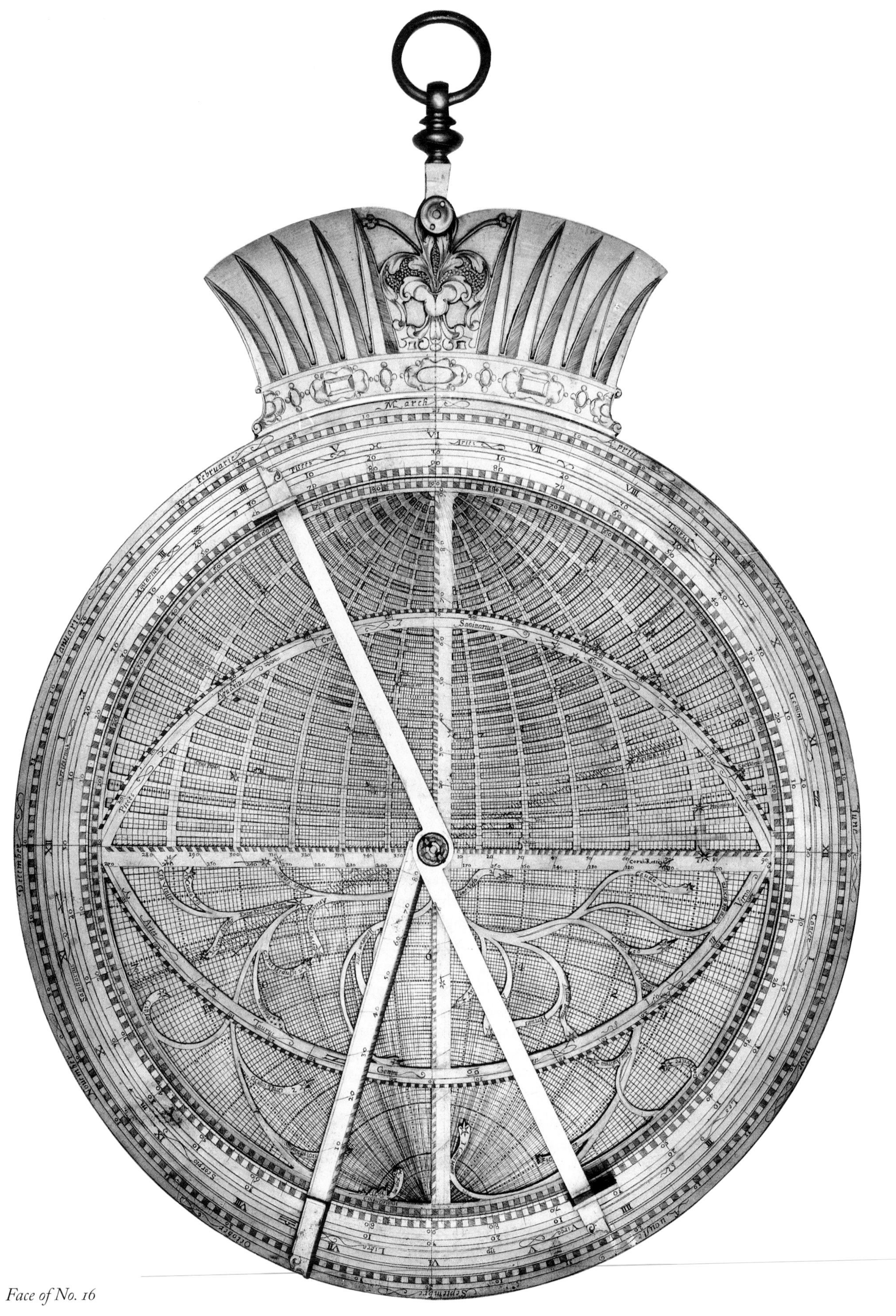

Face of No. 16

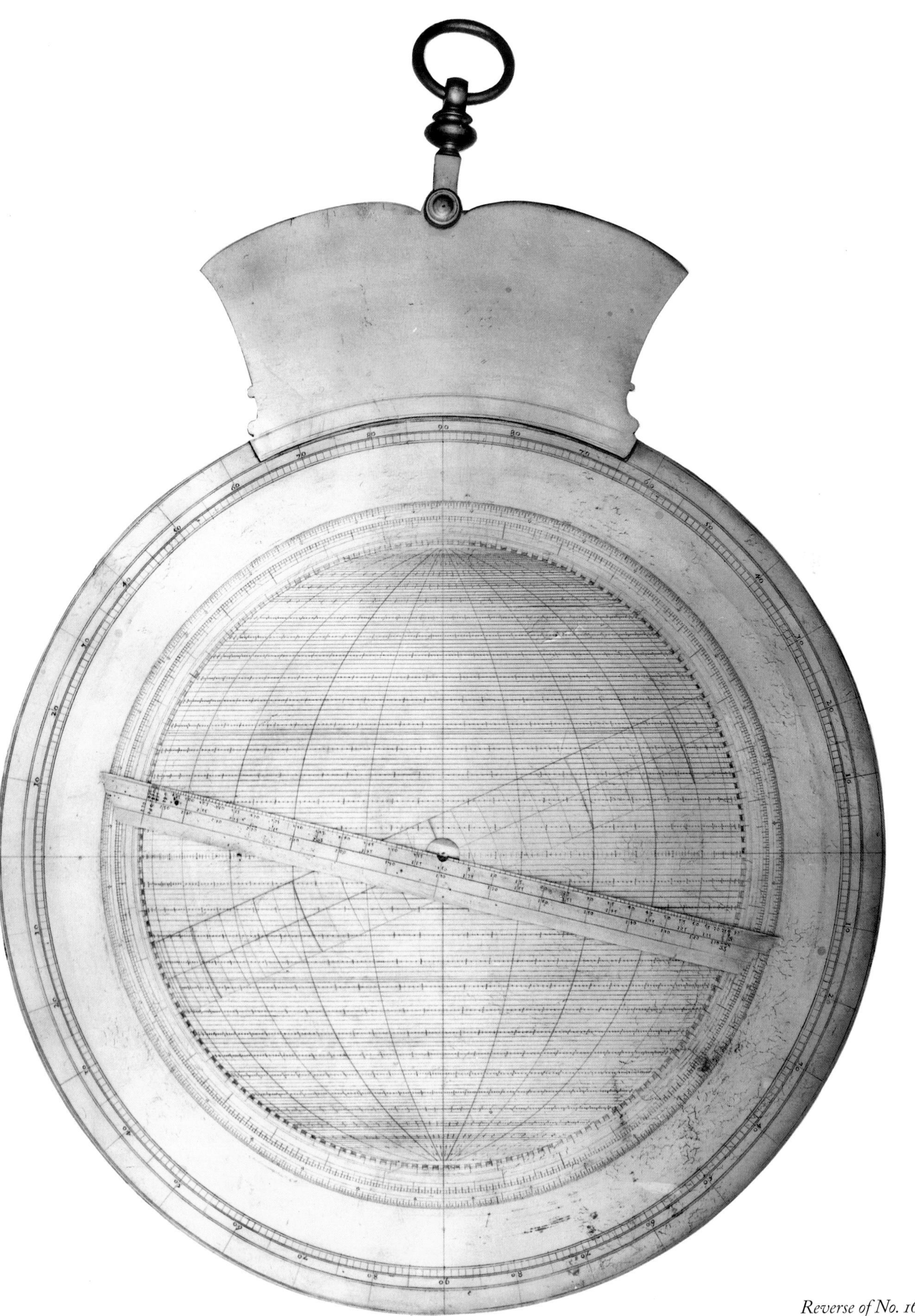

Reverse of No. 16

★ STAR LIST ON RETE

Star name	*Mag.*	*Modern name*
Corui ala dextra	3	Gamma Corvi
Cauda ♌		Beta Leonis
Hume: Vrsa	2	? Epsilon Ursae Majoris
Ceruix Leo		Theta Leonis
Cor Leonis		Alpha Leonis
Cor Hydre		Alpha Hydrae
Asellus Austrinu		Gamma Cancri
Asellus Boreus		Delta Cancri
Hercules	2	Beta Geminorum
Canis Minor	I	Alpha Canis Minoris
Apollo	2	Alpha Geminorum
Argo Naui		Alpha Carinae
Canis Maior	I	Alpha Canis Majoris
[unnamed]		?
hu: dex Ori		Alpha Orionis
Cingu Orionis		Zeta Orionis
Cin		Epsilon Orionis
Orio		Delta Orionis
[unnamed]		?
Sinister ps: Orionis		Beta Orionis
Hircus		Alpha Aurigae
Hadorum prece		Eta Aurigae
[unnamed]		[Zeta Aurigae]
Oculus Tauri	2	Alpha Tauri
Pleiadum [6 stars]		Pleiades
Andoromade: achr		? Gamma Andromedae
Caput Medulse	2	Beta Persei
Ceti Juba		Alpha Ceti
Cornu Arietis	2	Alpha Arietis
Extre: Eridani Acarn		Alpha Eridani
Cauda Ceti		Beta Ceti
Andro Scapu:		? Andromedae
[unnamed]		?
Ceti Venter		Zeta Ceti

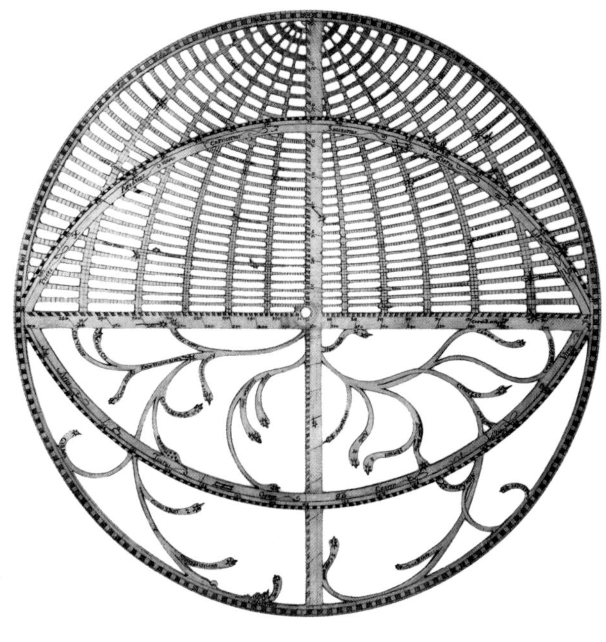

Rete of No. 16

with an extra decorated piece at the end. It is divided by degrees alternately hatched and marked and labeled 0° to 90° to the center. The marks match the scale on the prime meridian line. There is a wing nut and washer to fasten the astrolabe.

The universal astrolabe was invented in the eleventh century by 'Alī ibn Khalaf in Andulusia.[4] It appears in volume III of the thirteenth-century *Libros del saber,*[5] on a fourteenth-century astrolabe from Aleppo, now in the Benaki Museum in Athens,[6] and in the *Astronomicum caesareum* by Petrus Apianus.[7] The astrolabe in quadrant form at the Adler Planetarium, A-108 (cat. no. 36) and two quadrants by Christopher Schissler[8] carry the grid over the usual projection. The latter were in the Stadtmuseum in Kirchheim, although one has been transferred to the Stuttgart Württemburgische Landesmuseum.

There is a smaller version of the Blagrave rete, attributed to Abraham Sharp (1651-1742), at Bolling Hall near Bradford, England. There was also an astrolabe at the Basle Observatory, signed "B. A. 1561" (now lost), that carried a quarter grid on the reverse.[9] A manuscript in Madrid shows a similar rete.[10]

A similar astrolabe, signed "Carolus Whitwell," is in the Museo di Storia della Scienza in Florence. Robert Dudley is known to have brought instruments made by Whitwell (fl. 1590-1611) to Florence.[11] Gerard Turner thinks that M-33 may be by Whitwell.[12]

David A. King of the University of Frankfurt comments as follows:

"This fine Elizabethan universal astrolabe is a testimony to the unity of astronomical instruments in the Middle Ages and Renaissance. It shows that the foremost instrument-makers in sixteenth-century England were interested in the same problems as those in eleventh-century Toledo, where the universal astrolabe was invented, and that they favoured the

☆ STAR LIST ON GRID

Star name	*Modern name*
Crus Pegasi	Beta Pegasi
Fomahand	Alpha Piscis Australis
Cauda Capricor:	Gamma Capricorni
Dex hu Ceph	Alpha Cephei
Cauda Dephu	? Delphini
Cauda cigni	Alpha Cygni
Cuspis Sagitari	Epsilon or Lambda Sagittarii
Aquila	? Alpha Aquilae
Fidicula	Alpha Lyrae
[unnamed]	?
Caput Dra	Gamma Draconis
Caput Ophiuchi	Alpha Ophiuchi
Caput Engouna	Alpha Herculis
Palma Ophiuci	Delta Ophiuchi
Cor Scorpi	Alpha Scorpii
Aust	Gamma Scorpii
Borea	Xi Scorpii
Frons	Delta Scorpii
Lucida Corp: Gnos	Alpha Coronae Borealis
Lancus Borea	Beta Librae
[unnamed]	?
[unnamed]	?
Arctu Bootes	Alpha Bootis
Spica Virginis	Alpha Virginis
Ursa	? Alpha Ursae Majoris
Preuindemiat	Epsilon Virginis
Maio	? Ursae Majoris
Corui Rostř	Alpha Corvi

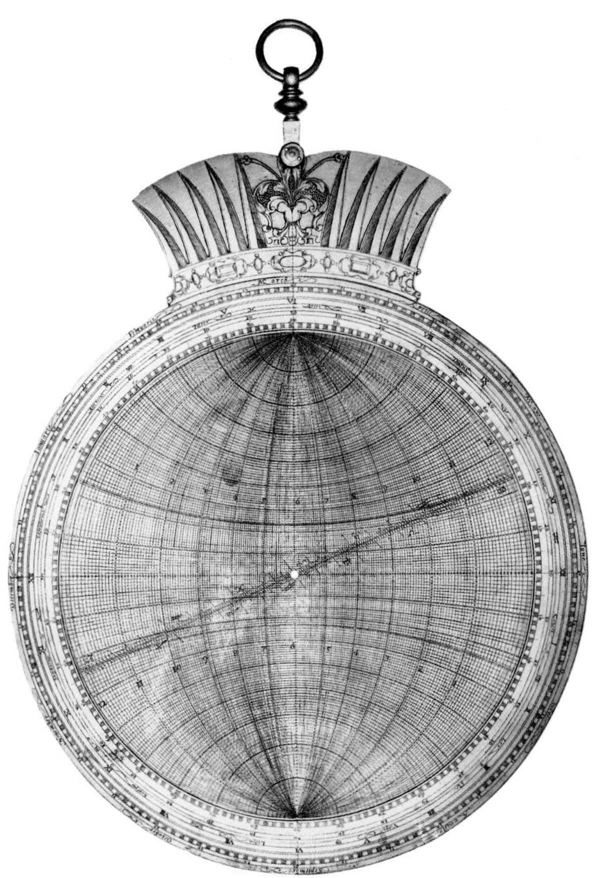

Mater of No. 16

same instrumental solution. How John Blagrave knew of the universal astrolabe has yet to be explained. There is evidence that this unsigned piece was made either by Charles Whitwell or by James Kynvyn, both working in London during the 1590s."

The Adler's instrument, M-33, which follows Blagrave's description, appears to have been made around 1620, based on the shift in the star positions due to precession.

For all its complexities of use, Blagrave's *Mathematical Jewel* persisted in favor for many years. Both William Blundeville[13] and John Palmer[14] wrote books to explain its workings.

DATE ACCORDING TO PRECESSION — *c.* 1620.
I ARIES = 21 March (Gregorian calendar).
PROVENANCE — A. W. M. Mensing, Amsterdam, 1924; Max Adler, Chicago, 1930; A. P. gift, 1930.
REFERENCES — Engelmann Catalogue (1924), 15, plates 1, 6; Fox (1932), 337-38, fig. 13; Fox (1933), 35, fig. 33 (p. 37); Gibbs *et al.* (1973), 15; Gunther (1932), 501, plate 141; King (1979a), plate 2; King (1979c), 244, plates 1, 3; Maddison (1966), appendix; North (1966), 64 n. 19; Price (1955), 255; Turner, A. J. (1973), 66-67; Turner, A. J. (1985), 160.

1. Blagrave (1585).
2. *Ibid.*, book 5: 75-92; North (1966), 64.
3. Blagrave (1585), plates 1, 2, 3.
4. King (1994), letter.
5. Rico y Sinobas (1863-68), 3, book 1: between 10 and 11.
6. King (1979a), plate 2; King (1979c), plate 5.
7. Apianus (1540), fol. M-4.
8. Dreier (1979), 93-96.
9. Zinner (1965), 149, 229, Tafel 50, no. 1.
10. King (1979c), plate 3.
11. Turner, G. L'E. (1991), 82.
12. Turner communication, 1997.
13. Blundeville (1594).
14. Palmer (1658).

17 de Rojas-type astrolabe

Germany?
Early 17th century
Unsigned
26.0 x 18.6 x 2.3 cm.
Rete thickness — 0.1-0.2 cm.
Brass
ICA 2001
M-42

Face of No. 17

The mater and throne are formed from a hammered sheet of brass and an annular ring laminated together. The throne is shield-shaped, with a round, empty hole in the center to take a press-fit compass, now missing. There is an extra ring surrounding the hole to give more depth for the compass box. The bail is pivoted and carries the ring.

The limb on the face has two scales. The outer one is divided by degrees and marked and labeled every ten degrees, 90°-0°-90°, twice from the top. The inner scale is divided every degree and alternately hatched. The two degree scales coincide with each other.

The de Rojas projection has eighteen named stars, numbered in the order of their Right Ascension. The meridians are drawn every five degrees, 66.5° to 66.5°, emphasized every fifteen degrees, and labeled "HORAE ANTE MERIDIEM" and 1 to 11 above the top line and "HORAE POST MERIDIEM" and 11 to 1 below.

The parallels are drawn between the Tropics, with every two intermediate ones shown as dotted lines. The double ecliptic connects the Tropics and carries the zodiacal signs, with Aries and Libra on the meridian

Reverse of No. 17

line at the center point. The spaces between 66.5° and 90° at the top are labeled "POLUS ARCTIC." and, at the bottom, "POLUS ANTARC."

The positions of the stars are indicated by asterisks (★).

The regula and cursor are missing. They have been replaced by a modern double counterchanged rule, which is incorrect.

The reverse of the astrolabe has several scales on the limb. The outer one is an hour scale, labeled XII to XII twice. Within this scale is a circle of degrees labeled every ten degrees, 90°-0°-90°, twice. The innermost scale is divided every degree and marked and labeled every ten degrees, clockwise, with 180° at the top and 360° at the bottom.

A rotatable disc fits into the cavity. It carries a double, concentric calendar scale with both the degree and the day scale alternately hatched. The zodiacal calendar has both signs and names engraved. The month names are in Latin, with "Mayus" and "Novēber" being the variant spellings. The innermost ring carries sixteen numbered asterisks, referring to the first sixteen stars on the de Rojas face, in order of their Right Ascension. Each is opposite the date on which it crosses the prime meridian at midnight, thus turning this disc into a "ready reckoner" (●).[1]

● READY RECKONER

1. [May 25]
2. [June 4]
3. [June 11]
4. [June 26]
5. [June 28]
6. [July 8]
7. [July 10]
8. [Aug. 5]
9. [Aug. 16]
10. [Aug. 24]
11. [Oct. 9]
12. [Oct. 17]
13. [Oct. 28]
14. [Nov. 25]
15. [Dec. 26]
16. [Jan. 1]

★ STAR LIST

No.	*Star name*	*Modern name*
1.	Oc. t.	Alpha Tauri
2.	Hir.	Alpha Aurigae
3.	D. h. Or.	Alpha Orionis
4.	Cano.	Alpha Carinae
5.	C. ma.	Alpha Canis Majoris
6.	C. ♊ ant.	Beta Geminorum
7.	C. m.	Alpha Canis Minoris
8.	Luc. hẏ.	Alpha Hydrae
9.	Cor L.	Alpha Leonis
10.	Cau. Le.	Beta Leonis
11.	Sp. v.	Alpha Virginis
12.	E. C. ur. mi.	Alpha Ursae Minoris
13.	Arct.	Alpha Bootis
14.	Cor sc.	Alpha Scorpii
15.	Lÿr.	Alpha Lyrae
16.	Aq.	Alpha Aquilae
17.	Cẏg.	Alpha Cygni
18.	Cau. Cap.	Delta Capricorni

Within the upper half of the circle of asterisks there is an equal/unequal hour scale for conversion of the hours according to the time of year. The equal hours are labeled 1 to 12 twice along the mid-line. The unequal hours are labeled 1 to 12 on alternate lines along the top arc. This type of conversion scale is often found on Habermel, Arsenius, and Gemini astrolabes.

The shadow square is divided on the outside into 60 parts and labeled by tens, 10-60-10, twice. On the inside it is divided into twelve parts and marked and labeled by fours, 4-8-12-8-4, twice.

The reverse of the disc shows an earlier attempt to engrave the face, but since the hour scale lines were drawn straight, a new start had to be made.

The alidade is counterchanged and has two scales. The left-hand one is labeled "Horae ortus et occa: ☉" It is divided and labeled 1 to 10 to the center. This scale is used in conjunction with the diagram of equal/unequal hours. The right-hand scale is marked for declination but is unusable with the de Rojas projection. It is marked every five degrees and labeled every ten, 0° to 70°, toward the center.

---

1 ARIES = 20.75 March, from the calendar scales.
PROVENANCE — A. W. M. Mensing, Amsterdam, 1924; Max Adler, Chicago, 1930; A. P. gift, 1930.
REFERENCES — Engelmann Catalogue (1924), 16; Gibbs *et al.* (1973), 23; Maddison (1966), appendix.

1. Saunders (1984), 70-71.

18 Classic-type astrolabe

Germany
1620
"Anno Domini 1620"
on the cross-bar of the rete
35.5 x 26.8 x 2.2 cm.
Rete thickness — 0.4 cm.
Copper
ICA 287
M-34

Face of No. 18

The copper rim of the mater is soldered to the hammered copper backplate. The throne shows a mermaid facing both ways and with a split tail, plus leafy scrolls. It is fastened to the mater by two bolts let into the mater. The bail and the ring are pivoted.

On the face of the mater there are two scales, the outer one being for hours, labeled 12 to 12 twice, with each hour divided into twelve parts, marked by threes. The inner scale is divided by degrees and marked and labeled every ten degrees, 0° to 360°, clockwise from the east.

The cavity is engraved as a tympan for 45°. The azimuths are drawn and labeled every ten degrees from 90° at the top to 0° at the horizons. The Tropics, the equator, and both horizons are labeled. The crepuscular line is labeled on the left "Crepusculum Matutinum" and on the right "Crepusculum Vespertinum." The Great Houses are marked and labeled from the left horizon, 1 to 12 counter-clockwise. The almucantars are labeled every four degrees from the top, 24° to 90°. Below the equator is inscribed "Duodecim Domicilia Coeli / Horae Planetarum." Around the edge of the tympan, from the top clockwise, "Meridies," "Occidens," "Septentrio," and "Oriens" are inscribed.

The reverse of the mater carries the two eccentric calendars, marked and labeled in the usual way. Both names and signs of the zodiac are shown. The months are in Latin. The eccentric space between the two calendars is inscribed "Perigaum ⨀" at the bottom. Below both calendars the other eccentric space carries "Apogaum ⨀" at the top. The mid-line is labeled "Medio Distancia ⨀" twice and carries the divisions and signs of the zodiac for use with the unequal hour scale. Above the mid-line, a double unequal hour scale is labeled outside, 1 to 12 clockwise. Two concentric circles lying between the unequal hour lines are marked

★ STAR LIST

Star name	*Modern name*
Marchab Pegasi	Alpha Pegasi
Scheat	Delta Aquarii
Clara D Humeri ♒	Epsilon Aquarii
Pectus Cygni	Gamma Cygni
Alchar Fidicula	Alpha Lyrae
Prima Antinoi	? Alpha Antinoi
nova Stella 1600	
Ras Aben	Gamma Draconis
Nova stellaA 1604	
Ras Alhague	Alpha Ophiuchi
Ras Algethi	Alpha Herculis
Alpheta	Alpha Coronae Borealis
Alramech	Alpha Bootis
Azimech	Alpha Virginis
Algorab	Gamma Corvi
Cometa 1618	
Alkes	Alpha Crateris
Kalb Eleced	Alpha Leonis
Alioth	Epsilon Ursae Majoris
Alphard	Alpha Hydrae
Algomeisa	Alpha Canis Minoris
Alhabor	Alpha Canis Majoris
Bed Algeuze	Alpha Orionis
Medium Leporis	Alpha Leporis
Alhejoth	Alpha Aurigae
Aldebaran	Alpha Tauri
Algol	Beta Persei
Stella Eridani	Alpha Eridani
Menkar	Alpha Ceti
Nova stella 1572	
Deneb Kaytos	Beta Ceti
Baten Kaytos	Zeta Ceti
Alpharez	Alpha Andromedae

Reverse of No. 18

"Horologium Veterum Hebraeorū Chaldaeorū Aegyptiorum" around the upper half, and "Horae Planet" is engraved under "Chaldaeorū." Inside the circles is a seven-pointed, shaded star with the planetary symbols at the tips, the solar symbol being at the top. The spaces in between the points are labeled 1, 3, 5, 7, 2, 4, 6 and run counter-clockwise from just to the right of the north point.

The shadow square is divided into twelve parts, marked and labeled by twos. Each vertical side is labeled "Vmbra Recta," while "Vmbra Versa" is under the square and "Scala Altimetra" is within it. Outside the shadow square "Oriens" is inscribed on the left, "Occidens" on the right, and "Calendariū Gregorianum" and "Media Nox" across the base.

The rete shows 29 stars, the comet of 1618, and the novas of 1572, 1600, and 1604 (★). The star positions are shown by six-pointed, cut-out stars, engraved on the strapwork or attached by flame points. The comet is depicted with a tail and labeled "Cometa 1618." The novas are labeled "Nova stella 1572," "nova Stella 1600," and "Nova stellaA 1604." Kepler was much interested in the 1604 nova. The pattern is of interlocking arcs. The ecliptic circle is divided as usual, with both zodiacal names and symbols.

The rule is counterchanged and double. The alidade is counterchanged with folding sights.

DATE ACCORDING TO PRECESSION — *c.* 1614.
I ARIES = 19.5 March, from the calendar scales.
PROVENANCE — A. W. M. Mensing, Amsterdam, 1924; Max Adler, Chicago, 1930; A. P. gift, 1930.
REFERENCES — Engelmann Catalogue (1924), 15, plate 1; Fox (1932), 338, fig. 14; Fox (1933), 37, fig. 34; Gibbs *et al.* (1973), 15; Gunther (1932), 460; Price (1955), 256, 259; Schechner Genuth (1997), 106.

19 Classic-type astrolabe

Philippe Danfrie and Jehan Moreau
Paris
1584 and 1622
"Philipus Danfrieus Siderographus Regius / Generalis Lutetiae exarabat Anno / Salutis 1584"
below the shadow square
"A PARIS Chez Jehan / Moreau Rue St. / Jacques a la Croix blan / che / 1622."
on the back, at the top and between the calendars
31.0 x 21.8 x 6.0 cm.
Rete thickness — 0.19 cm.
Wood, paper, and brass
W-98

Face of No. 19

The mater is of wood covered with paper printed from an engraved plate. The throne is trilobed, with the face of a woman on the front and a satyr on the back. The pivoting bail and ring are of brass. The limb carries the hour scale, labeled XII to XII twice. It is divided by degrees and labeled by fives, 90°-0°-90°, from the top. The cavity has a slot for tympan tangs. The inner rim is faced with paper.

On the reverse of the mater, the limb carries a circle of degrees, divided every half degree and labeled 90°-0°-90° twice from the top. The next scale is divided by degrees, alternately hatched and labeled by fives, 5° to 30°, twelve times. This scale is used for the zodiacal calendar and is labeled with the appropriate pictured signs and names. This is followed by the Gregorian calendar, which is mounted eccentrically and shows the epacts. The names of the months are in Latin, with "Janvarius," "Junius," and "Julius" being the variant spellings. The allegorical symbols for the months are engraved.

In the upper half of the center section are two circular calendar tables with the address of Moreau in between, at the top. In the lower left corner of the left quadrant is inscribed: "Hoc Dorsum Kalendario / reformata precise / acōmodatum / est" (This back is precisely adjusted to the reformed calendar).[1]

On the left circle, the outer scale has "1583-1586" printed at the bottom, followed by two manuscript dates, "1627" and "1635." The next scale is labeled "Cyclus Solaris" and numbered 1 to 28 counter-clockwise. The next two scales are related. The first is labeled "LITER" and carries four sets of the dominical letters, each with one or two letters missing. The last scale is labeled "DOMINICAL" and carries the missing letters, appropriately placed. In the center is inscribed: "Usus harum duarū / Rotularum ab anno. / 1582 usque in ultimum / diem futuri proxime /

◆ INDEX PASCHAE

25 mr	3 ap	
5 ap	14 ap	[Opposite "1585" in outer ring]
16 ap	26 mr	[Opposite "1584" in outer ring]
28 mr	6 ap	[Opposite "1583" in outer ring]
8 ap	17 ap	
18 ap	29 mr	
31 mr	9 ap	
11 ap	23 mr	
23 mr	1 ap	
	12 ap	

■ EPACTAE

xxi	xxix
viii	xviii
xxvii	vii
xvi	xxvi
v	xv
xxiiii	iiii
xiii	xxiii
ii	xii
xxi	i
x	

Reverse of No. 19

Seculi duraturus est." (These two wheels can be used from 1582 until the last day of the next century.)

The right-hand calendar scale's outer ring is labeled "1583, 1584, 1585" near the bottom. The next scale is labeled "INDEX PASCHAE," with a list of dates running clockwise (◆).

The following circle is labeled "EPACTAE," with Roman numerals running clockwise (■).

The innermost circle is labeled "Aure numer," with 1 to 19 running clockwise. In the center is inscribed: "Die dominica pxime / Sequenti Pascatis / In dicem qui est / Luna decima quarta / Primi mensie sacrū / Pascha celebratr." (Easter is celebrated on the first Sunday following the Paschal Index, which is the full moon of the first month.)

The shadow square shows two scales. The inner one is divided into twelve parts and labeled by threes. The outer one is divided into 60 parts, labeled by fives. The Danfrie inscription is below the square. Within the square is inscribed: "Ad inuendienda autem, nouilunia Epactae / Iuxta singulos dies mensium hic scriptae / sunt Quare cuicunq diei adiacet ānî labentis / Epacta eo die fieri nouiluniū scire conuenit." (The epacts are written next to the days of the month so that one can find the new moon. Thus, the epact for the current year adjacent to a given date indicates a day of new moon.)

Both of these statements refer to the new Gregorian calendar, the determination of Easter, and the epacts. The usual labels of "Vmbra Versa" and "Vmbra Recta" are shown.

The rete has 28 flame-shaped star points (★) and is of paper on cardboard. The pattern is interlocking scrolls. The ecliptic is divided by degrees and labeled with zodiacal names and signs. The reverse is plain.

There are four tympans, made of paper mounted on cardboard. They are for 48° (with the other side left blank), 42° and 54°, 45° and 51°, and

★ STAR LIST

Star name	Mag.	Modern name
CRVS PEGASI	2	Beta Pegasi
LIRA	1	Alpha Lyrae
CAVDA SIGNI	2	Alpha Cygni
CAVDA CAPRICOR	3	Delta Capricorni
CAVDA DELPHIN	3	Epsilon Delphini
AQVILA	2	Alpha Aquilae
CAPVT OPHIVCHI	3	Alpha Ophiuchi
COR SCORPII	2	Alpha Scorpii
PALMA OPHIVCHI	3	Delta Ophiuchi
ARCTVRVS	1	Alpha Bootis
EXTRE CAV VR MA	2	Eta Ursae Majoris
SPICA VIRGINIS	1	Alpha Virginis
PRIMA CAVDE VR MA	2	Epsilon Ursae Majoris
CAVDA ♌	2	Beta Leonis
ROSTRVM CORVI	3	Alpha Corvi
COR LEONIS	1	Alpha Leonis
LVCIDA HIDRAE	2	Alpha Hydrae
CANIS MINOR	1	Alpha Canis Minoris
CANIS MAIOR	1	Alpha Canis Majoris
DEX HVMERVS ORIO	1	Alpha Orionis
HIRCVS	1	Alpha Aurigae
SINISTER PES ORION	1	Beta Orionis
OCVLVS TAVRI	1	Alpha Tauri
DEX LAT PER	2	Alpha Persei
CAP MED	2	Beta Persei
VENTER CETI	3	Zeta Ceti
ANDROMEDA	3	Beta Andromedae
CAVDA CETI	3	Beta Ceti

Rete of No. 19

39° and 57°. Each has the usual lines drawn, labeled "Horizon Obliqus," "Horizon Rectus," "Tropicus Cancri," and "Capricorni," as well as "Linea Aurorae siue Crepusculinae" and "Aequinoctialis." The Great Houses and unequal hours are also shown.

The alidade, which is made of brass, is counterchanged and plain. The rule, also brass, is double and counterchanged. The ends are pointed and have a leafy decoration. The bolt and wing nut are original.

Danfrie engraved the plates for his astrolabe in 1578 and, after having corrected the civil calendar, reissued the instrument in 1584. After his death in 1606, the engraving plates went through several hands. In 1622 Jehan Moreau added his name and address to the plates and reissued the paper astrolabe. He also planned to engrave new, up-to-date plates in 1625, but no further record is known.[2]

This astrolabe is identical to ICA 2007 at the Smithsonian Institution's National Museum of American History. ICA 2007 was sold at Sotheby's on February 26, 1962, and is fully described by Gibbs with Saliba.[3]

Minor conservation was done on W-98 by Graphics Conservation, Inc., Chicago, in 1983.

I ARIES = 21 March, from the calendar scales.

PROVENANCE — Paris dealer, 1984; R. S. Webster, Chicago, 1984; A. P. gift, 1984.

REFERENCE — Turner, A. J. (1989), 32.

1. This and the following three translations are from Gibbs with Saliba (1984), with the kind permission of the Smithsonian Institution Press.
2. Turner, A. J. (1989), 32.
3. Gibbs with Saliba (1984), 16, 25, 41-42, 154-56, figs. 27, 101.

20 Classic-type astrolabe

Germany?
c. 1600
Unsigned
36.3 x 27.6 x 3.6 cm.
Rete thickness — 0.5 cm.
Brass
ICA 531
M-29

Face of No. 20

This astrolabe is unfinished. The rete is the only part completely divided but only partially labeled, as the star points are not named. The mater consists of a cast rim with the backplate soldered to it. The throne is decorated with three rosettes with small foliate scrolls on either side. It is pinned and soldered to the mater and has a pivoted bail to hold the suspension ring.

The face of the mater is incomplete. The outer scale is divided by degrees and is further marked to indicate a double 12-to-12 hour scale. The space within this scale is unmarked. The cavity is unmarked except for the meridian and center lines, and it has neither slots nor lugs to hold the tympans.

The reverse of the mater has the zodiacal calendar divided every degree and marked every five degrees and 30 degrees. The next scale is divided into 73 parts, each representing five days. The two inner spaces are unmarked. The center area is blank, except for the meridian and the right horizon.

The rete carries 24 six-pointed stars, each with a long pointer to indicate the position of the star. None are named (★). The ecliptic circle is divided by degrees, marked and labeled every five degrees, and identified every 30 degrees with the

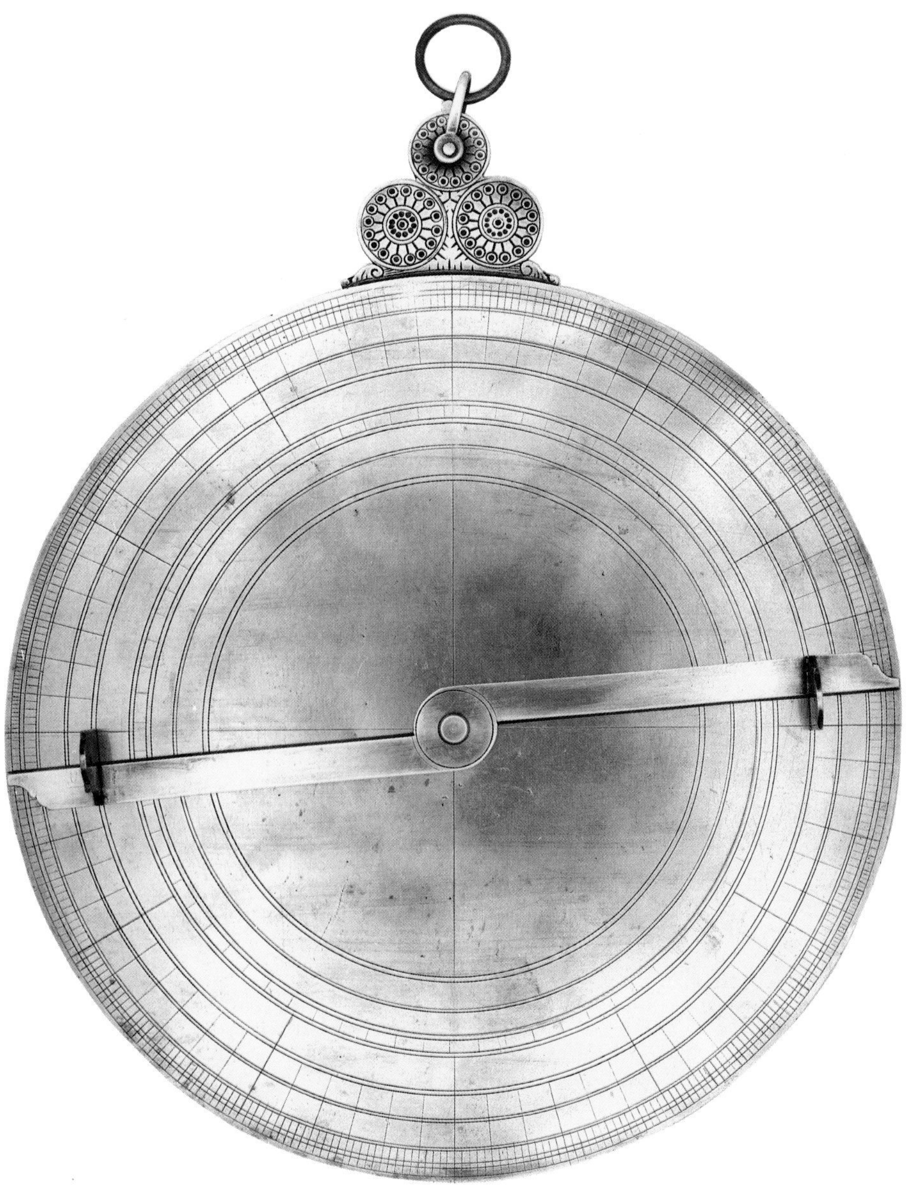

STAR LIST

- Alpha Pegasi
- Delta Aquarii
- Epsilon Pegasi
- Delta Capricorni
- Alpha Aquilae
- Alpha Lyrae
- Alpha Ophiuchi
- Alpha Scorpii
- Zeta Ophiuchi
- Alpha Coronae Borealis
- Alpha Bootis
- Alpha Virginis
- Epsilon Ursae Majoris
- Beta Leonis
- Alpha Hydrae
- Alpha Cancri
- Alpha Canis Minoris
- Alpha Canis Majoris
- Beta Orionis
- Alpha Tauri
- Beta Persei
- Zeta Ceti
- Beta Andromedae
- Alpha Cassiopeiae

Reverse of No. 20

zodiacal symbol. The celestial equator is shown on the rete, divided by degrees and marked and labeled every five degrees, 0° to 360°, from the first point of Aries counter-clockwise. The reverse of the rete shows construction lines.

There is only one tympan, with the face drawn for 45° latitude. The almucantars are five degrees apart, the azimuths ten degrees. The usual lines are shown, plus the Great Houses and the unequal hour scale. The center points for the various almucantars are visible. There are no labels. The reverse is blank.

The alidade is counterchanged, and the ends are cut away in curves. The rule is double and counter-changed, and the ends are finished like the alidade. The bolt and wing nut appear to be original.

The rete of this unfinished astrolabe is carefully laid out, with only the star points unlabeled. The tympan may be a later addition. It is very unusual to have the almucantar center points punched so as to be noticeable.

DATE ACCORDING TO PRECESSION — *c.* 1602.
PROVENANCE — A. W. M. Mensing, Amsterdam, 1924; Max Adler, Chicago, 1930; A. P. gift, 1930.
REFERENCES — Engelmann Catalogue (1924), 14, plates 1, 6; Gibbs *et al.* (1973), 18; Price (1955), 257.

21 Multiple-type astrolabe

South Germany
c. 1620
Unsigned
16.5 x 11.7 x 2.1 cm.
Rete thickness — 0.5 cm.
Brass
ICA 532
M-32

The mater, including the throne, was sawn from one piece of brass, and the cavity was turned out on a lathe. There is a straight edge at the bottom for use with a plane table. The throne is a cut-out A-shaped scroll with a hole in the center for the knob of the swivel-bail. The ring pivots and swivels also. Little floral rosettes decorate the ends of the scroll. The cavity is empty except for a notch to accommodate the tympan tang.

The limb has an hour scale, labeled XII to XII twice and divided into 360 degrees, alternately hatched, and marked and labeled every ten degrees, 90°-0°-90°, twice from the top clockwise.

On the reverse side, the limb carries two concentric calendar scales, on which the degrees and days are alternately hatched. The names of the zodiacal signs are given, and the months are shown in Latin, with "Mayus" being the only variant spelling. The unequal hour scale is double and labeled 1 to 12 counter-clockwise. There is a large, double rosette in the center.

The shadow square is divided into twelve parts alternately hatched, marked and labeled by threes, 3-12-3, twice. It is appropriately labeled "VMBRA VERSA" and "VMBRA RECTA."

The rete has nine named stars with flame-shaped star points (★). The reverse is plain.

The single tympan is silvered and has the tang at the top. The face is for 45° latitude and shows the Great Houses, labeled on the outer limit from the left horizon, 1 to 12 counter-clockwise. The equator and the Tropics are also delineated. The almucantars are drawn every five degrees and labeled by fives, 5° to 20° on each side, from the horizon, and also on the meridian.

The reverse of the tympan carries an Azarquiel or Gemma Frisius projection. The line of the ecliptic is divided and labeled with the signs of

★ STAR LIST

Star name	*Modern name*
CAVDA SIGNI	Alpha Cygni
SPICA VIRGINIS	Alpha Virginis
ARTVRVS	Alpha Bootis
COR LEONIS	Alpha Leonis
CANIS MINOR	Alpha Canis Minoris
CANIS MAIOR	Alpha Canis Majoris
PES SINISTER ORIONIS	Beta Orionis
OCVLVS TAVRI	Alpha Tauri
VENTER CETI	Zeta Ceti

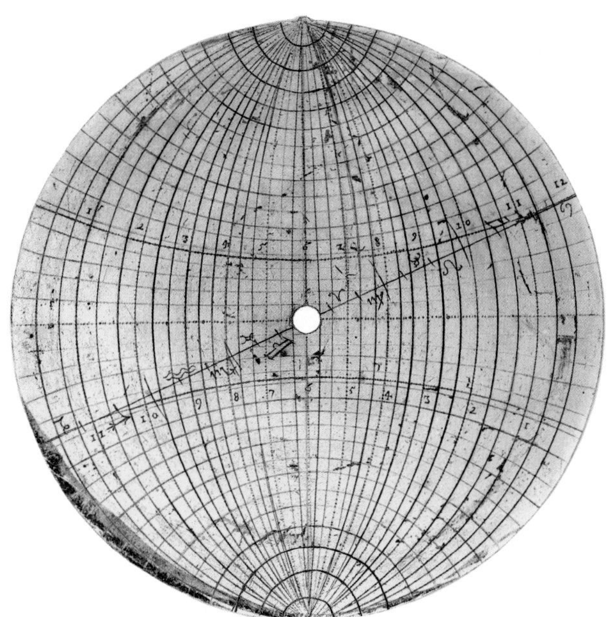

OPPOSITE, LEFT: *Face of No. 21*
OPPOSITE, RIGHT: *Reverse of No. 21*
ABOVE: *Tympan of No. 21, side with Azarquiel projection*

the zodiac. The meridians are labeled along the Tropic of Cancer 1 to 12 clockwise, and 12 to 1 clockwise along the Tropic of Capricorn. No stars are shown.

The alidade is counterchanged and unmarked, with tapered ends. The regula is unmarked, with fluted ends. It is dovetailed to carry the missing cursor with brachiolus. It carries a post with a slot for the horse. It also serves as a double rule when the rete is in use. All parts appear to be original.

This astrolabe has many similarities with Leiden A-553,[1] including the overall form and the layout of the rete. It is described in the ICA as being signed "SABEI," made of paper, and dated *c.* 1590, but these statements are all in error.[2]

DATE ACCORDING TO PRECESSION — *c.* 1620.
1 ARIES = 21 March, from the calendar scales.
PROVENANCE — A. W. M. Mensing, Amsterdam, 1924; Max Adler, Chicago, 1930; A. P. gift, 1930.
REFERENCES — Engelmann Catalogue (1924), 15, plates 1, 6; Gent (1994), 41; Gibbs *et al.* (1973), 18; Gunther (1932), 364, plate 87; Price (1955), 255.

1. Van Cittert (1954).
2. Gibbs *et al.* (1973), 18.

22 Double planispheric astrolabe

Isaac Habrecht 2
Strasbourg, France
c. 1628
"Planigloby Coelestis pars australis / Inventore isaaco habrecto argentino"[1]
along the lower edge of Side A
36.0 x 36.0 x 0.3 cm.
Manuscript on cardboard
A-304A

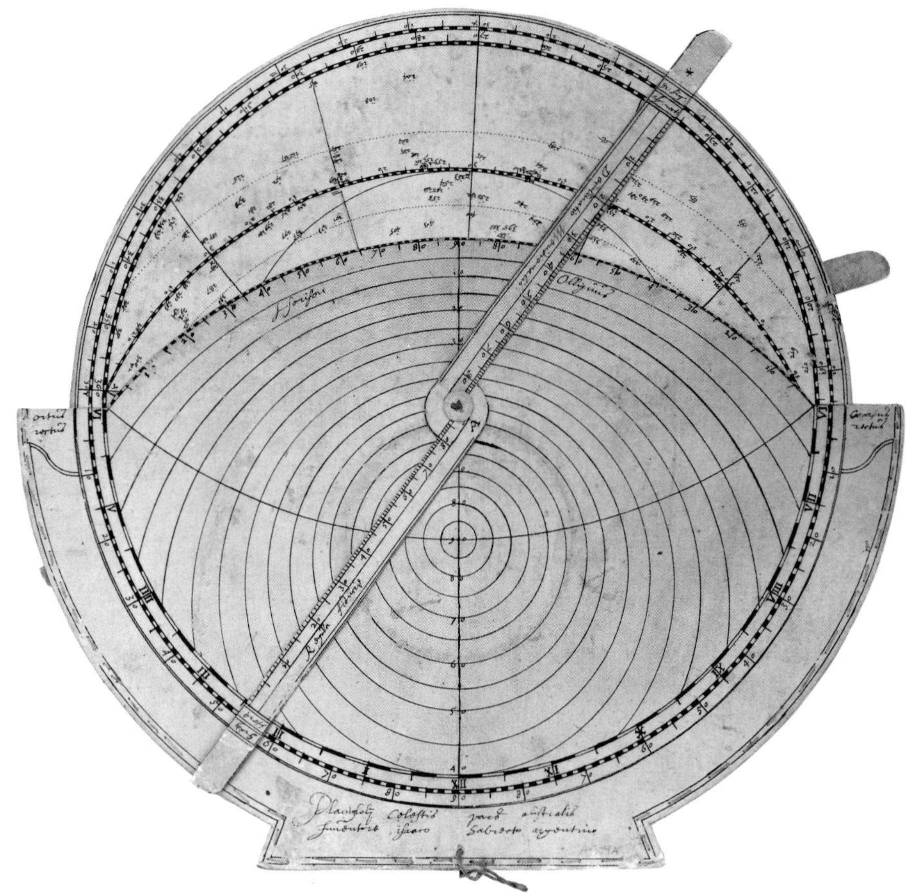

Side A of No. 22

This cardboard manuscript astrolabe shows the stars on a double-faced, revolving disc, sandwiched between the front and back of the mater. Each face has the celestial equator as the outer limit, thus making a complete sphere. It is laid out for 50° latitude. Side "A" uses the North Celestial Pole as the center of the projection, while Side "B" uses the South Celestial Pole, the more usual projection.[2]

The side of the mater labeled "A" on the rule shows the lines of equal altitude, which are drawn every five degrees and labeled every ten on the meridian, 40°-90°-0°, at the top. The upper edge of the convex half of the mater is labeled "Horizon Obliquus" and divided by degrees unequally, with alternate black and white spaces. It is labeled every ten degrees, 0°-90°-0°. The mater also carries two scales. The outer one shows degrees and is labeled like the horizon. The inner one is labeled VI-XII-VI clockwise and divided into fifteen-minute intervals. The left end of the horizon line, or east, is labeled "Ortus rectus." The right end, or west, reads "Occasus rectus." Along the lower edge of the mater is inscribed: "Planigloby Coelestis pars australis / Inventore isaaco habrecto argentino."

Both halves of the counterchanged rule are divided by degrees and labeled every ten degrees, 0° to 90°, toward the center. One half is labeled "Declinatio Meridionalis" under the scale. Its outer end is labeled "Gr: Sig:" and under that "afr: -et:" for use with the scales on the disc. The other half is labeled "Regula Fiducis" under the scale and, at the end, "Grady" with "Horol:" under it. These are for use with the scales on the mater.

The back of the mater, labeled "B" on the rule, also has the lines of equal altitude labeled 40° to 0° along the meridian and divided as on Side A. The "Horizon Obliquus" is concave on this side and is similarily divided and labeled. The left end,

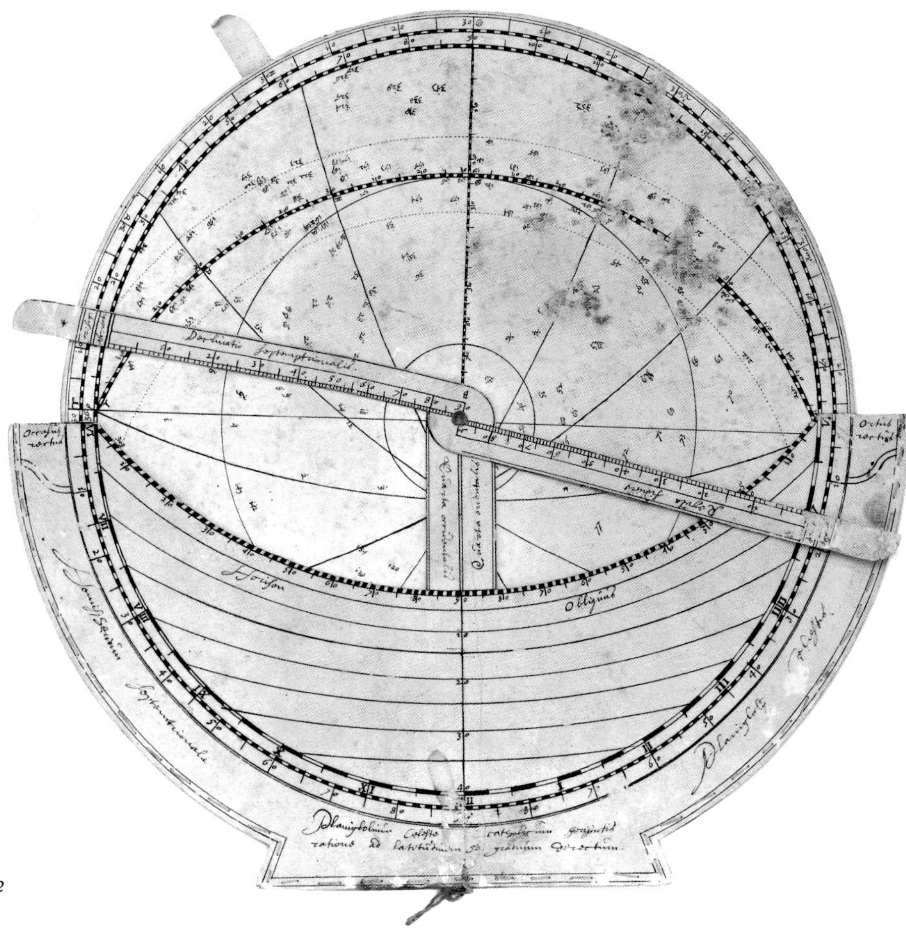

Side B of No. 22

west, is marked "Occasus rectus," and the right end, east, is marked "Ortus rectus."

The mater is labeled "Hemispherium Septentrionale" and "Planigloby Coelestis." The bottom of the mater is inscribed: "Planiglobium Coelestis ___[?] ___[?] / Ratione ad latitudinum 50 gradium ___[?]."

The counterchanged rule on Side B is divided and labeled as on Side A. One half is marked "Declinatio Septemptrionalis" and terminates with the inscriptions "Gr: sig:" and "asc ___[?]" for use with the scales on the disc. The other half is labeled "Regula fiducis," and the end is labeled as on Side A for use with the scale on the mater. The rule is carried on a wide bar to raise it above the mater. One section is labeled "Quārta occidentalis" and the other "Quārta orientalis."

The star chart is double-sided. Side A shows the half of the ecliptic that contains ♈, ♓, ♒, ♑, ♐, ♏, ♎ clockwise. The stars are represented by numbered dots, running from 83 to 411. Side B shows the other half of the ecliptic, containing ♈, ♉, ♊, ♋, ♌, ♍, ♎ clockwise. The stars are numbered from 1 to 356. The star numbers on each side are independent. Thus, as an example, for the numbers between 300 and 309, one side has 300 to 302 and 305 to 308, while the other side has 303, 304, and 309. There is no repetition from one side to the other. All the ones are dotted. Unfortunately, the star list is missing. The mid-line of the ecliptic is divided by degrees and marked and labeled by tens, 90° to 0°, from the top, although on Side A only 40° to 0° shows. All scales are done in alternating white and black squares.

This star disc is, in some ways, quite similar to that in M-31 (cat. no. 15).

Habrecht's *Planiglobium coeleste, et terrestre* was first published in Strasbourg in 1628.

PROVENANCE — American dealer, 1994; A. P. purchase, 1994.

1. F. R. Maddison correspondence, June 1994; Dr. Francis Debeauvais correspondence, Nov. 1994 and Jan. 1995.
2. Morrison (1994).

TOP LEFT: *Side B (Boreale), showing the northern sky, from Habrecht (1666)*

TOP RIGHT: *Side A (Australe), showing the southern sky, from Habrecht (1666)*

RIGHT: *Double title page, from Habrecht (1666). Courtesy of the University of Chicago Library.*

PLANIGLOBIUM
COELESTE,
Hoc est
GLOBUS COELESTIS
NOVA FORMA AC NORMA IN PLANUM PROJECTUS, OMNES ORBIS cœlestis lineas, circulos, gradus, partes, stellas, sidera &c. in planis tabulis æri incisis artificiosè exhibens.

ADJECTA SUCCINCTA TUM FABRICÆ tum usus explicatione, omnium Problematum, quæ vulgatis hactenus globis, Planisphæriis, Astrolabiis expediri solita sunt, facilimam solutionum continente.

UNA CUM ANNEXA METHODO SUB FINEM, qua quilibet ope hujus Instrumenti, sine ullo manuductore, aut vivo præceptore, stellas cœli quascunq; cognoscere & denominare possit.

ISAACI HABRECHTI.
Phil. & Med. Doct.
PLANIGLOBIUM
COELESTE AC TERRESTRE
Argentorati quondam;
Nunc
Operâ
JOHANNIS CHRISTOPHORI STURMII
Norimbergæ,
Emendatiùs, auctiùs ac universaliùs editum,
Prostat apud PAULUM Fürsten / Techno-bibliopolam Norimbergensem.

Typis CHRISTOPHORI GERHARDI.

23 Astrolabe, partial

[Isaac Habrecht 2]
[Strasbourg, France]
c. 1628
Unsigned
36.5 x 30.0 x 0.25 cm.
Manuscript on cardboard
A-304B

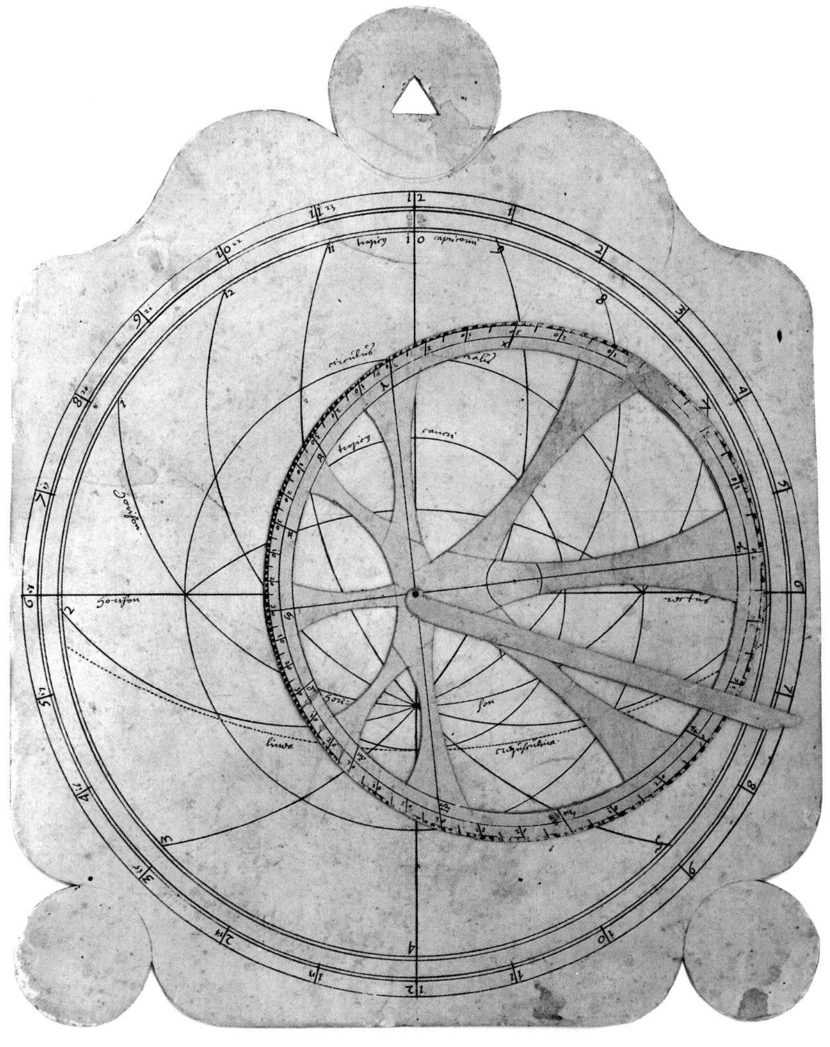

Face of No. 23

The mater is essentially rectangular with some decorative cut-outs. The throne is part of the mater. The limb has an hour scale on it, marked and labeled in hours, 12 to 12 twice. The second twelve hours are also numbered 13 to 24. The mater has the lines for a tympan inscribed on it. It shows the Tropics of Capricorn and Cancer, the equator, the right and oblique horizons (all in Latin), and the Great Houses, which are numbered counter-clockwise, starting at the oblique horizon.

The rete consists only of the ecliptic circle supported by a number of spokes. It is divided every degree and labeled 10, 20, 30 twelve times with the appropriate zodiacal symbols. It has a single, unmarked rule. The reverse of the instrument is blank.

This instrument is surely by Isaac Habrecht 2, the maker of A-304A (cat. no. 22). A further piece of ephemera by the same maker, A-304C, came with A-304A and B. It shows the start of a layout of an astrolabe on one side and a diagram of the "Systema Tychonicus," with various tables and scales, on the other. Dr. Francis Debeauvais has kindly drawn our attention to another example, *c.* 1665, in brass, in the Museo Settala in the Pinacoteca Ambrosiana in Milan.

PROVENANCE — American dealer, 1994; A. P. purchase, 1994.

24 Classic-type astrolabe

Italy
c. 1650
Unsigned
33.0 x 24.5 x 4.5 cm.
Rete thickness — 0.8 cm.
Brass
ICA 179
M-30

Face of No. 24

The mater consists of a cast ring with the backplate soldered on to it. A magnetic compass is inset in the face of the throne, which has scrolling ends and a leafy pattern. The compass rose is labeled "ME," "OC," "SE," and "OR." The needle, the mica glazing, and the spring-retainer are original. A swivel and suspension ring are present. The throne is riveted to the mater. The back of the throne has a floral design.

The rim has an hour scale, labeled XII to XII twice. Each hour is divided into quarters. A degree scale, lying inside the hour scale, is marked every five degrees and labeled every ten degrees, 90°-0°-90°, twice from the top.

The cavity has a decorative set of crosslines and circles engraved in it. There is a lug at the top to hold the tympans.

On the reverse of the mater the usual double calendar scales are shown. They are eccentric. The names and signs of the zodiac are both shown. The civil calendar has Latin month names. Above the center line is a double unequal hour scale, labeled 1 to 12 counter-clockwise. The shadow square below the center line is labeled "VMBRA VE:" twice and "VMBRA RECTA" along the

Reverse of No. 24

base line. Each of the four sides is divided into 50 parts, marked by tens and labeled every 20 parts, 20 to 100.

The rete carries 32 flame-shaped star points (★). The names and astrological signs are both given, as are some magnitudes. On the back of the rete the ecliptic circle is completely divided and marked every ten degrees and 30 degrees.

There are four tympans, all of which have a notch at the top to accommodate the lug in the mater's cavity. The usual lines are present, labeled "TROPICVS CANCRI," "TROPICVS CAPRICORNI," and "AEQVINOCTIALIS." The "LINEA CREPVSCVLI" is shown and labeled. The unequal hour lines are shown below the horizons and labeled either I to XII or 1 to 12. Tympans #3 and #4 also show the Great Houses. The tympans are drawn for latitudes of 38° and 39°, 40° and 41°, 42° and 43°, and 44° and 45°. Some show more almucantars and azimuths than others.

The alidade has notched sights and cut-out pointers. It is counter-changed and blank. The rule is single and unmarked, and its pointer is similar to the alidade's. There is a bolt with a wing nut and two washers. All parts appear to be original. The washers may indicate one or more missing tympans.

★ STAR LIST

Star name	*Mag.*	*Modern name*
Dex. hum. Pega	2	Beta Pegasi
Crus Aquary	3	Delta Aquarii
Pectus Pegasi	3	Epsilon Pegasi
Caud cigni	2	Alpha Cygni
Aquila Volans	2	Alpha Aquilae
Rostr galinae	3	Beta Cygni
Prima Antinoi	3	Alpha Antinoi
Cap Herculis	3	Alpha Herculis
Ophiuci Dex. genu	3	Eta Ophiuchi
COR SCORPI	1	Alpha Scorpii
Sinist genu [Ophiuci]	3	Zeta Ophiuchi
Lucida lancis Aust.	2	Gamma Scorpii
Arcturus	2	Alpha Bootis
Vlt. caud. Vrsae		Eta Ursae Majoris
Spica ♍		Alpha Virginis
Cauda ♌		Beta Leonis
In Basi Crateris	4	? Crateris
Dorsum ♌	2	Delta Leonis
Cor ♌		Alpha Leonis
Lucida Hydrae	1	Alpha Hydrae
Canis minor	1	Alpha Canis Minoris
Canis maior	1	Alpha Canis Majoris
media Cing Orio	2	Epsilon Orionis
Hirc^9 Capel	1	Alpha Aurigae
Cornu ♉ Aust:	4	Zeta Tauri
Ocul^9 Tauri	1	Alpha Tauri
Post interualum fluuy	4	Theta Eridani
Dex. lat^9 Persei	4	Alpha Persei
Cap. Algol	2	Beta Persei
Cing. Andromedae	3	Beta Andromedae
Cauda Ceti Aust:	3	Beta Ceti
Pect^9 Cassiop.		Alpha Cassiopeiae

The symbol "9" is the Gothic shorthand for the ending "us."

TOP: *Tympan #1 of No. 24, left for 38°, right for 39°*
BOTTOM: *Tympan #2 of No. 24, left for 40°, right for 41°*

This astrolabe, with its magnetic compass, could be used as a surveying instrument.

DATE ACCORDING TO PRECESSION — *c.* 1653.
1 ARIES = 21 March, from the calendar scales.
PROVENANCE — A. W. M. Mensing, Amsterdam, 1924; Max Adler, Chicago, 1930; A. P. gift, 1930.
REFERENCES — Engelmann Catalogue (1924), 14, plates 1, 6; Gibbs *et al.* (1973), 14; Gunther (1932), 331, plate 73; Price (1955), 253.

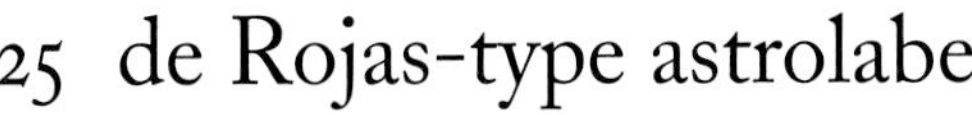

25 de Rojas-type astrolabe

Pierre Sevin
Paris
1682
"P. SEVIN A PARIS / 1682"
on the reverse, at the bottom of the de Rojas disc
34.6 x 26.6 x 0.85 cm.
Rete thickness — 0.45 cm.
Brass
ICA 2080
M-463

MONTHS

Ian	Iuil.
Feb	Aoust
Mars	Sept.
Avril	Octo.
May	Noue.
Iuin	Decem.

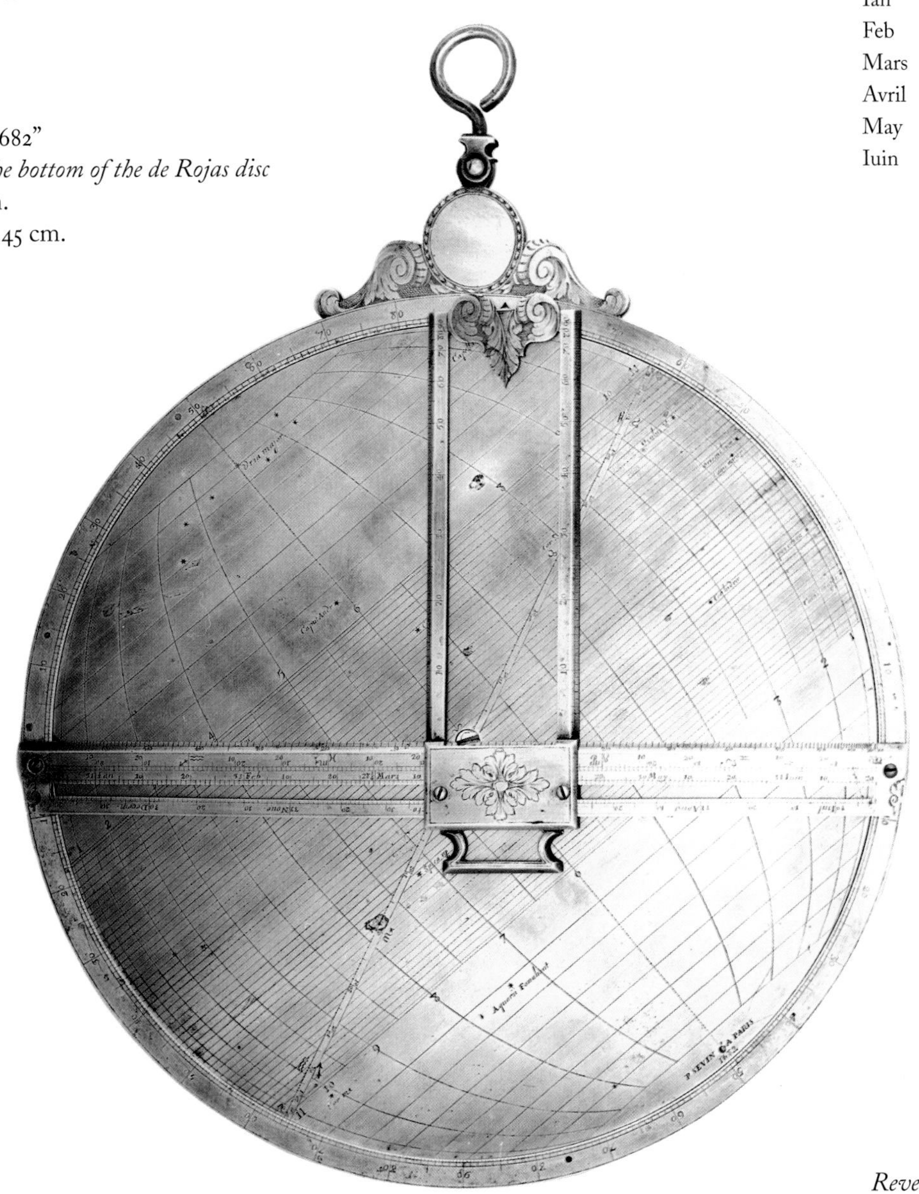

Reverse of No. 25

The mater consists of two rings riveted to a central disc, forming a cavity on both sides. The throne has two facing scrolls on either side of a circular disc, which has a blank center surrounded by a leaf border. The bail pivots and the ring swivels.

The face carries an outer scale that is divided every degree. The next scale is divided every half degree, marked every five and ten degrees, and labeled 90°-0°-90° twice from the top. The hour scale is labeled XII to XII twice, at fifteen-degree or fifteen-minute intervals from the top.

The empty cavity is shallow, with no slot or pin to hold a tympan. It probably held a "ready reckoner,"[1] which would have shown the radial list of stars from the de Rojas projection in order of their Right Ascension, the dates on which they would have crossed the prime meridian at midnight, and their declination (see picture of M-465B, cat. no. 27). A decorative disc was added as a filler at some later date.

The reverse has a rotatable disc with a de Rojas projection and a fixed regula with a sliding cursor. The limb has a scale of degrees, with the first, fifth, and tenth marked and labeled every ten degrees, 90°-0°-90°, twice from the top.

The de Rojas projection shows 20 stars (★). Between the Tropics, the parallels of declination are drawn every degree. Above and below the Tropics, they are drawn every five degrees to 66.5°. Between the Tropics, 23 meridians are drawn. The prime meridian goes from pole to pole. Beyond the Tropics eleven meridians are shown. Above the Tropic of Cancer they are labeled I to II, and below Capricorn, II to I. The line of the ecliptic is shown as a band to make the labeling easier. The upper edge is the significant one. It is divided every degree by pricks, marked every five degrees, and labeled every ten, 10°, 20°, 30°, six times from the center, which represents Aries and Libra. Each

★ STAR LIST

Star name	*Modern name*
Ursa major	Eta Ursae Majoris
"	Zeta Ursae Majoris
"	Epsilon Ursae Majoris
"	Delta Ursae Majoris
"	Gamma Ursae Majoris
"	Beta Ursae Majoris
"	Alpha Ursae Majoris
Cauda ♌	Beta Leonis
Cor ♌	Alpha Leonis
Cor hidrae	Alpha Hydrae
Oculus ♉	Alpha Tauri
Capella	Alpha Aurigae
Orionis hum	Alpha Orionis
Canis minor	Alpha Canis Minoris
pes orion	Beta Orionis
Canis major	Alpha Canis Majoris
Caput Andr	Alpha Andromedae
Aquarii Fomahant	Alpha Piscis Australis
Cor ♏	Alpha Scorpii
Lira	Alpha Lyrae

Mater of No. 25

30° space is marked appropriately with two signs of the zodiac. The south polar area is engraved: "P. SEVIN A PARIS / 1682." There is a small hole between the "N" and the "A." The reverse is blank.

The regula on the reverse has a beveled edge and a slot along its length in which the cursor slides. It is fixed to the limb with two screws and carries four scales. The one on the beveled edge is divided unequally and labeled 90°-0°-90°, with 0° at the center part. The next scale is the zodiacal calendar, which is divided every degree, marked every five and ten degrees, and labeled 10°, 20°, 30° twelve times, six across and six back. The zodiacal signs are also labeled. The last two scales are the civil calendar, running on either side of the slot. Each scale is divided by days, marked by fives and tens, and labeled 1 to 28¼, 30, or 31 as appropriate, with the months' names in French (■). The upper scale runs from December 21 to June 20, and the lower scale runs back from June 21 to December 20.

The cursor, which runs on the regula, has two separated scales, with a pointer at the top that has a foliate design. The cursor has a foliate design and a decorative bracket below. Each vertical bar carries the same scale, 0° to 90°, from the regula to the top; they are unevenly divided. There are two bolts that hold the cursor together.

DATE ACCORDING TO PRECESSION — *c.* 1690.
I ARIES = 20 March, from the calendar scales on the regula.
PROVENANCE — A. W. M. Mensing, Amsterdam, 1924; Max Adler, Chicago, 1930; A. P. gift, 1930.
REFERENCES — Fox (1933), 35; Gibbs *et al.* (1973), 24; Gibbs with Saliba (1984), 224; Maddison (1966), appendix.

1. Saunders (1984), 70-71.

26 de Rojas-type astrolabe

Germany
c. 1690
Unsigned
19.2 x 13.2 x 1.7 cm.
Rete thickness — 0.3 cm.
Brass
ICA 2081
M-465A

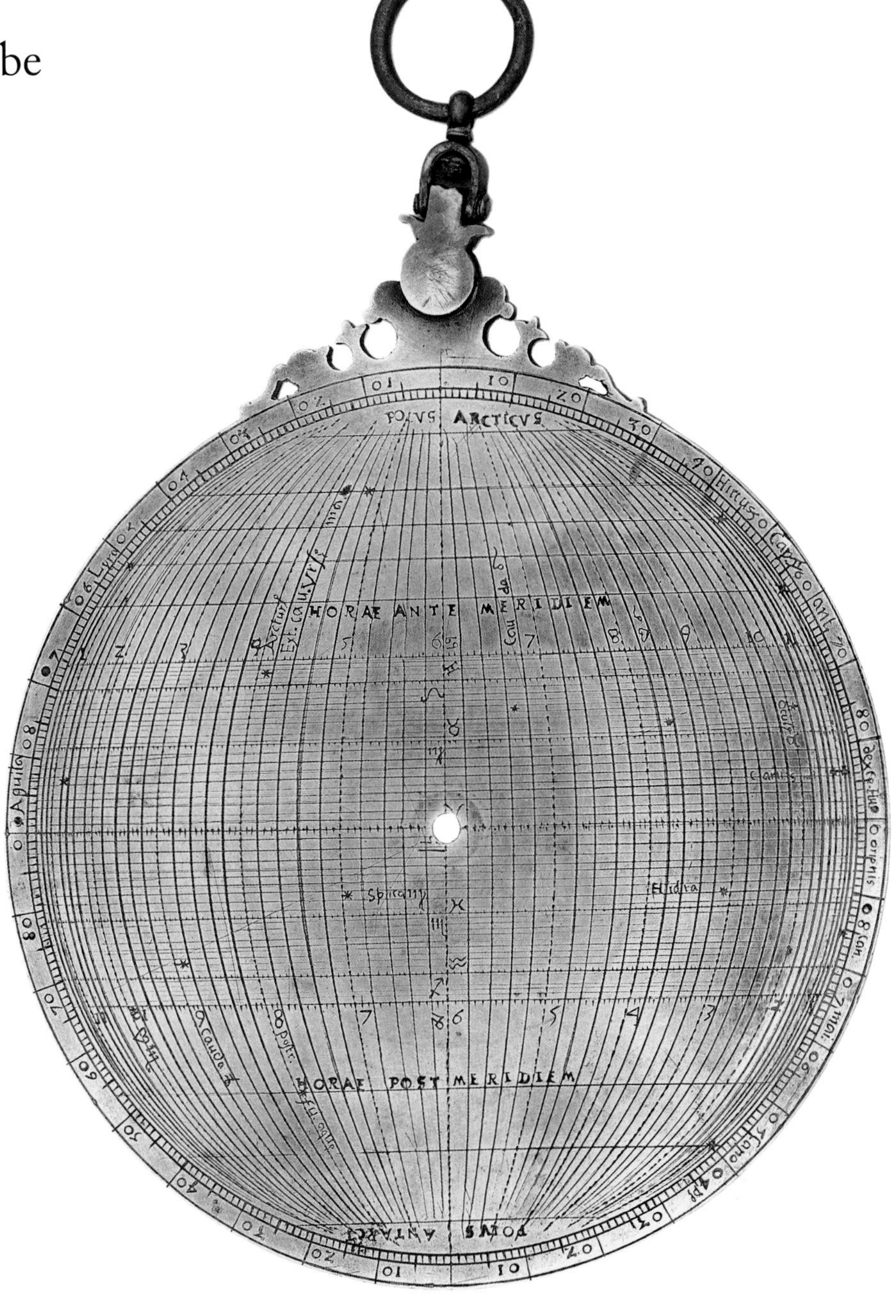

Reverse of No. 26

The mater is made of two brass plates soldered together. The cavity was formed by cutting out the center section of the face plate before lamination. The throne, which was sawn out earlier, is triangular and cut out in a decorative fashion. It carries a pivoting bail and ring attachment. The swiveling ring is present.

The face has a scale on the limb, divided by degrees, marked every fifteen degrees, and labeled 12 to 12 twice from the top. The scale is also marked every five and ten degrees and labeled every ten degrees, 90°-0°-90°, twice from the top on the inside. The cavity is empty and blank. It probably held a "ready reckoner" such as M-465B (cat. no. 27), but it has a different star list and its own hour scale.

The reverse carries a de Rojas projection. The limb is divided by degrees, marked every five and ten degrees and labeled 0°-90°-0° twice from the top. The parallels are divided every degree between the Tropics and every ten degrees above and below, and the Tropics are pricked. Beyond the Tropics, to 23.5°, the parallels are drawn every ten degrees. The meridians are drawn every three degrees, and every fifth one is pricked. They sloppily extend into the polar zones. The prime meridian is pricked and carries the

Face of No. 26

★ STAR LIST

Star name	*Modern name*
Ext. Cau. Vrs^{9} mn.	Alpha Ursae Minoris
Hircus	Alpha Aurigae
Lyra	Alpha Lyrae
Cap. ♊ ant.	Alpha Geminorum
Arctur^{9}	Alpha Bootis
Cauda ♌	Beta Leonis
Ocul^{9} ♉	Alpha Tauri
♡ ♌	Alpha Leonis
Canis mi	Alpha Canis Minoris
Dexte. Hu. orionis	Alpha Orionis
Aguila	Alpha Aquilae
Spica ♍	Alpha Virginis
Hidra	Alpha Hydrae
Can. mai.	Alpha Canis Majoris
Cauda ♑	Delta Capricorni
♡ ♏	Alpha Scorpii
Postr. fu. aque	Alpha Piscis Australis
Canop^{9}	Alpha Carinae

The symbol "^{9}" is the Gothic shorthand for the ending "us."

signs of the zodiac, with Aries and Libra at the center. The poles are labeled "POLVS ARCTICVS" and "POLVS ANTARCT" (inverted). Above the Tropic of Cancer and below the Tropic of Capricorn are "HORAE ANTE MERIDIEM" and "HORAE POST MERIDIEM," respectively. Across the Tropic of Cancer the hours are labeled 1 to 11, and below Capricorn they are labeled in reverse order, with 12 being crowded out in both instances. The line of the ecliptic is lightly drawn in. Eighteen stars are shown (★).

The regula and sliding cursor to be used with the de Rojas projection are missing.

The twos are shown as "z"s.

DATE ACCORDING TO PRECESSION — *c.* 1690.
PROVENANCE — A. W. M. Mensing, Amsterdam, 1924; Max Adler, Chicago, 1930; A. P. gift, 1930.
REFERENCES — Fox (1933), 35; Gibbs *et al.* (1973), 24.

27 "Ready reckoner" from a de Rojas-type astrolabe

c. 1700
Unsigned
11.8 x 0.1 cm.
Brass
ICA 2081
M-465B

★ STAR LIST

Star position		*Modern name*	*Date position*
Eridan	C	Alpha Eridani	[May 7.5]
Ocul ♉	7	Alpha Tauri	[May 27]
Hircus	3	Alpha Aurigae	[June 4]
pes si. Or.	A	Beta Orionis	[June 5]
Hu. dex Or	10	Alpha Orionis	[June 17]
Canop	D	Alpha Carinae	[June 26]
can mi	9	Alpha Canis Minoris	[June 29]
Sirius	B	Alpha Canis Majoris	[July 13]
Cor ♌	6	Alpha Leonis	[Aug. 16]
Cau ♌	5	Beta Leonis	[Sept. 15]
Spic ♍	E	Alpha Virginis	[Oct. 11]
Vlt ca Vr	I	Eta Ursae Majoris	[Oct. 18]
Arctur ·	4	Alpha Bootis	[Oct. 25]
Cor ♏		Alpha Scorpii	[Nov. 26]
lira	2	Alpha Lyrae	[Dec. 27]
Aquila	8	Alpha Aquilae	[Jan. 11]
Aqua	F	Alpha Piscis Australis	[Feb. 24]

Ready reckoner of No. 27

This "ready reckoner" disc[1] is part of a de Rojas-projection astrolabe that has been lost. The back is blank, but the face carries several scales. The first is an hour scale, divided every 20 minutes and labeled 12 to 12 twice. The second, a zodiac calendar, is divided every degree, marked every five degrees, and labeled with the zodiacal signs. The last scale, a civil calendar, is divided into 365 days, marked every fifth or sixth day as appropriate, and labeled with the months' initials, with "I" representing January, June, and July. The calendars are concentric.

The innermost circle contains a radial list of seventeen stars, six lettered A to F, ten numbered 1 to 10, and one unnumbered. They are listed opposite their appropriate dates of crossing the meridian at midnight. The stars are listed starting from the first point of Aries (★).

This disc definitely does not belong to M-465A (cat. no. 26), although it fits into the cavity. The star list is not identical and where the same star is listed, the form often varies. M-465C, a perpetual calendar with a volvelle, was included with M-465A and B. These are classic examples of odd pairings, marriages that happened to fit, which were not unusual for Mr. Mensing's workshop.

1 ARIES = 21 March, from the calendar scales.

PROVENANCE — A. W. M. Mensing, Amsterdam, 1924; Max Adler, Chicago, 1930; A. P. gift, 1930.

REFERENCES — Fox (1933), 35; Gibbs *et al.* (1973), 24.

1. Saunders (1984), 70-71.

28 Classic-type astrolabe

Reproduction of an instrument said to have been made by Piervincenzo Danti, now at the Museum für Kunst und Gewerbe, Hamburg
19th-century reproduction
"ALPHENOS SEVERVS GENIO SVO: ET COMMODITATI·F·"
between the calendars
"DIA ROMES. MED. V. CLI. / LAT. GR. XLII"
on the tympan
48.6 x 27.0 x 3.9 cm.
Rete thickness — 0.2 cm.
Overall width — 36.2 cm.
Copper, silver, and copper gilt
DPW-51

Face of No. 28

The mater consists of a silver ring, or hoop, to which is attached a gilded copper annular ring. This ring carries a 24-hour scale on the face, labeled I to XXIII, and a circle of degrees. On the back of the ring is a pin to locate the tympan. Because this astrolabe is hollow, it was necessary to have three radial pins protrude from the inside of the hoop to hold the tympan to the front. Notches in the tympan allow for its removal.

The throne is formed by two silver dolphins surmounted by a bearded satyr's head with a leaf cap. The mater is framed all around with an ornate stylized leaf pattern, made of silver, with silver-gilt finials and rosettes. At the bottom there are two gilded rosettes with a silver spike between them. The latter carries a punchmark, "ΛOL." This has been identified as a sixteenth-century mark from Aquila.[1]

The gilded copper backplate is a pressfit against a shoulder on the silver hoop. The inside is plain and ungilded. The reverse has two eccentric calendar scales. The months are in Latin, with "Februar" being the only variant spelling. On the zodiacal calendar, "Sagittar" is the only variant. The shadow square is divided into twelve parts and labeled III to XII four times. The inscription is set in a ribbon below the shadow square and between the two calendars. There is a coat of arms above the shadow square, inside the unequal hour scale. The latter is labeled I-VI-VI-I.

The rete is of silver and is formed by two intertwined snakes plus two pairs of satyrs' legs at Libra and Aries. There are 29 star points, all of which are named except for Cor Scorpii (★).

The tympan has gilding only on the face. There is a hole for the pin on the back of the annular ring. The face is inscribed: "DIA ROMES. MED. V. CLI. / LAT. GR. XLII." The usual lines are shown, with the

Reverse of No. 28

★ STAR LIST

Star name	*Modern name*
HV. DEX. EQ.	Beta Pegasi
EQVI ALA	Gamma Pegasi
CRVS	Delta Aquarii
CAVDA GALLINAE	Alpha Cygni
HVM. SI.	Epsilon Cygni
VVLTVR. VOL.	Alpha Aquilae
VVLT CA	Epsilon Aquilae
OPHIVCH	Alpha Ophiuchi
[Image of scorpion on apple in mouth]	Alpha Scorpii
CORONA	Alpha Coronae Borealis
LANX	Alpha Librae
BOOTES	Alpha Bootis
CAVD. VRS MA	Epsilon Ursae Majoris
CAVD. LEON	Beta Leonis
ALA D. CORV.	Gamma Corvi
DOR. L.	Gamma Leonis
VMB. VRS. MA	? Ursae Majoris
REX [within a ♡]	Alpha Leonis
HYDRA	? Alpha Hydrae
CANICVLA	Alpha Canis Minoris
SYRIVS	Alpha Canis Majoris
HVM. D. ORION	Alpha Orionis
ORION	? Orionis
H. OCVLVS	Alpha Tauri
PERSEI DEX.	Alpha Persei
C. GORG.	Beta Persei
OS. CETI	Gamma Ceti
PECT. CASSIOP	Alpha Cassiopeiae
VENT. CETI	Zeta Ceti
VMBILIC ANDRON	Alpha Andromedae
CAVD CETI	Beta Ceti

almucantars being drawn every three degrees and labeled every fifteen degrees. The unequal hour scale is labeled I to XII clockwise. There is a trial engraving for 42° on the reverse, which is not gilded.

The alidade is silver-gilded, counterchanged, and with two folding sights. The rule is silver and single. It is marked from the equator line outward 0° to 23.5° and labeled every ten degrees and also "Merid." It is marked from the equator line inward every other degree to the center and labeled X to LXII and "Septemtr."

This astrolabe is a reproduction of one at the Museum für Kunst und Gewerbe in Hamburg. The latter was in Frédéric Spitzer's collection and appeared in his catalogue[2] before being acquired by the museum in 1893.[3] In 1923 Rohde[4] ascribed it to Piervincenzo Danti of Perugia. Gunther, in 1932, followed Rohde's lead and stated that the coat of arms belonged to the Alfani family.[5] Alfano Alfani was Piervincenzo Danti's mathematics teacher prior to 1488. The Hamburg instrument is said to be a fifteenth-century work.

DPW-51 came with a wooden plaque as a filler. An attached paper noted that the astrolabe pertained to Galileo. There was also a small loan label: "Loan [printed] / 147.52 [in ink] / (8)."

Clare Vincent of the Department of European Sculpture and Decorative Arts, Metropolitan Museum of Art, has kindly furnished the following comments:

"A comparison of photographs of the Adler Planetarium's astrolabe with illustrations of the astrolabe in the Museum für Kunst und Gewerbe in Hamburg reveals that both the engraving of the mater and the rete and the ornament of the rete of the Adler astrolabe are moderately skillful copies of the engraving and ornament of the Hamburg astrolabe, but very much coarser in the execution of the details. In addition, the Adler's astrolabe has an ornamental frame not found on the Hamburg astrolabe. The throne of the Adler Planetarium's astrolabe especially is a travesty of the Hamburg throne, and it has been incorporated into a decorative design made up of scroll, floral, and palmette ornament, which, together with the throne, forms the whole circular frame attached to the mater. Further, a spike at the bottom of the frame, and integral to it, appears to have been created with the idea that the astrolabe could be set into some sort of supporting base or pedestal, although the attachment of the spike to the frame is, in fact, too flimsy for such a purpose. These features are fanciful conceits; nothing comparable is known to exist among the products of serious astrolabe makers.

"Considering the nature and the quality of execution of the Adler Planetarium's astrolabe, I would suggest that it was made at some time in the second half of the nineteenth century, perhaps in 1875, when a receipt for the Hamburg astrolabe, now in the Biblioteca Nationale Centrale in Florence, discovered by Prof. Thomas Settle and discussed by Francis R. Maddison in the exhibition catalogue, *Circa 1492*,[6] indicates that the Hamburg museum's astrolabe was in the temporary possession of the Military Engineers of Rome. It is also possible that the Adler Planetarium's astrolabe could have been made a few years later when the Hamburg museum astrolabe was in the possession of Frédéric Spitzer in Paris.

"The Hamburg astrolabe, itself, is unique in fifteenth-century Europe. While its ornamental motifs can be found in the ornamental vocabulary of the period, their use on the Hamburg astrolabe is not coherent in the way one would expect of a design of the period. The two elongated dragons with snake-like tails are unmistakably related to the serpentine motifs of the rete of the astrolabe attributed to Egnatio Danti in the Museo di Storia della Scienza in Florence.[7] Either the serpentine forms on Egnatio's rete are abstractions of his grandfather's design, or the maker of the Hamburg astrolabe worked from the design of Egnatio's rete. The second possibility is not inconceivable, for elements of the design of the Hamburg astrolabe, such as the palmettes, the hairy paws, Medusa head, and cornucopias with ribbon streamers, are motifs repeatedly employed by neoclassical designers, whose work reflects their familiarity with the ornaments found in the Roman ruins of Pompeii, excavated in the second half of the eighteenth century. The ornament of the rete and the engraved ornament on the reverse of the mater of the Hamburg astrolabe are not incompatible with a late eighteenth- or early nineteenth-century date although the astrolabe has been accepted as a product of the late fifteenth century.[8]

"There can be no question, however, that Alfano Alfani of Perugia did commission an astrolabe from Piervincenzo Danti de' Rinaldi, for it is mentioned as early as 1498 in the dedication of Piervincenzo's translation of Joannes de Sacrobosco's *La Sfera*. The 1579 edition of the translation published by Piervincenzo's grandson, Egnatio, reproduced the passage referring to the astrolabe in which Piervincenzo writes the following: '...I have resumed working on your astrolabe; in two months time I hope to have finished it with the greatest diligence I can muster; as you will see, I have marked the astronomical hours as you advised me to, and not in the ordinary way. Finally, I commend myself to you with great affection. Given at the Villa of Prepo on September 6, 1498.'[9] Whether the astrolabe in the Museum für Kunst und Gewerbe is wholly or partly later in date seems open to question."

1 ARIES = 11 March, from the calendar scales.

PROVENANCE — New York dealer, before 1953; David P. Wheatland, Topsfield, Mass., 1953; A. P. gift, 1986.

1. Information in letter from dealer to David P. Wheatland, 1953.
2. Spitzer Catalogue (1892), 2: 207, lot 2934.
3. Maddison (1991), 224-25.
4. Rohde (1923), 90-94.
5. Gunther (1932), 322-25.
6. Maddison (1991), 225.
7. Miniati Catalogue (1991), 44-45, fig. 27.
8. See, for example, Rohde (1923), 92-93, 90 Abb. 115, 91 Abb. 116; Gunther (1932), 2: 322-25, no. 171; *Circa 1492* Catalogue (1991), 224-25, no. 123; or Levi-Donati (1993), 79-107.
9. Sacrobosco (1579), TA: "...Ho rimesso le mani al vostro Astrolabio, e spero fra due mesi di tempo hauerlo condotto al fine con quella maggior diligenza che per me si potrà, oue vedrete, che ho segnate l'hore all' Astronomica come m'auuisaste, & non all' ordinario, con che facendo fine a voi con affetto di cuore mi raccomando. Dalla villa di Prepo all' 6. di Settembre nel 1498. Vostro molto amoreuole, Dante de' Rinaldi."

29 Classic-type astrolabe

Vibrandi
1595 (modern)
"VIBRANDI FEC. / 1595"
below the shadow square
16.0 x 11.6 x 2.2 cm.
Rete thickness — 0.3 cm.
Copper-gilded only on visible surfaces
ICA 2004
A-III

★ STAR LIST

Star name	*Modern name*
PV CAS	Alpha Cassiopeiae
CIG	Alpha Cygni
PIRM ♌	? Leonis
CAPRICORNVS AV	Gamma Capricorni
AQVILA	Alpha Aquilae
LIBR SC	Alpha Librae
DR	Alpha Draconis
FIDICVLA	Alpha Lyrae
SELI	? Beta Lyrae
RAE	?
HIR	? Herculis
LCO	Alpha Coronae Borealis
ELC	Delta Ursae Majoris
COB ♏	Alpha Scorpii
ILIVM	Epsilon Ursae Majoris
ART	Alpha Bootis
SCORVI ALA	Gamma Corvi
SPICA VIRGO	Alpha Virginis
LEONIS	Alpha Leonis
LVCIDA IDRAE	Alpha Hydrae
CANIS MINOR	Alpha Canis Minoris
SIRIVS COR	Alpha Canis Majoris
HIRCVS CA	Alpha Aurigae
PROC. MED. OR. I	Alpha Orionis
ALDEBARAN TAVRVS	Alpha Tauri
SIN. PES. ORIONIS RIGELA	Beta Orionis
VENTER CETI	Zeta Ceti

LEFT: *Face of No. 29*
RIGHT: *Reverse of No. 29*

The mater is of rolled gilt copper. The rim is soldered to the backplate. The throne has three rosettes with a shield, with the top rosette being a pivoting bail with ring. The face of the limb carries the hour scale, labeled XII to XII twice, with hatching and the usual divisions. The cavity is plain and has a notch for the tympan.

The reverse carries the two concentric calendars. The zodiacal names are used and the months are in Latin, with "Januarius," "Marcius," "Junius," and "Julius" being the variant spellings. There is a double unequal hour scale, labeled 1 to 12 from the left. The shadow square is divided into twelve parts and labeled by threes with the usual inscriptions. The signature and date, "VIBRANDI FEC. / 1595," are shown below the square.

The rete has 27 named star points, many of which are located incorrectly (★).

It is quite evident that the rete of this modern astrolabe was copied from a Bos astrolabe. By comparing the rete of A-III with that of M-33A (cat. no. 14), we have been able to identify which stars are shown, far-fetched as some of the names are. In the ecliptic circle, the signs for Scorpio and Virgo have been transposed.

The tympan, which is of rolled copper plate, is labeled "G. 37" on one side and is plain on the reverse. The almucantars are drawn and labeled every three degrees. The alidade and rule are present.

I ARIES = 12 March, from the calendar scales.
PROVENANCE — London dealer, 1941; A. P. purchase, 1941.
REFERENCE — Gibbs *et al.* (1973), 23.

30 Stellar compass, la Hire-projection astrolabe

Edward Ayearst Reeves
London
After 1908
"REEVES PATENT ASTRONOMICAL COMPASS NO. 113"
on the limb on the reverse
"REEVES' ASTRONOMICAL COMPASS PATENT APPLIED FOR NO. 14008.08 PUBLISHED BY EDWARD STANFORD, 12, 13, & 14 LONG ACRE, LONDON"
on the reverse
21.0 x 19.4 x 3.5 cm.
Rete thickness — 1.0 cm.
Aluminum, cardboard, plastic, and brass; wooden box
A-291

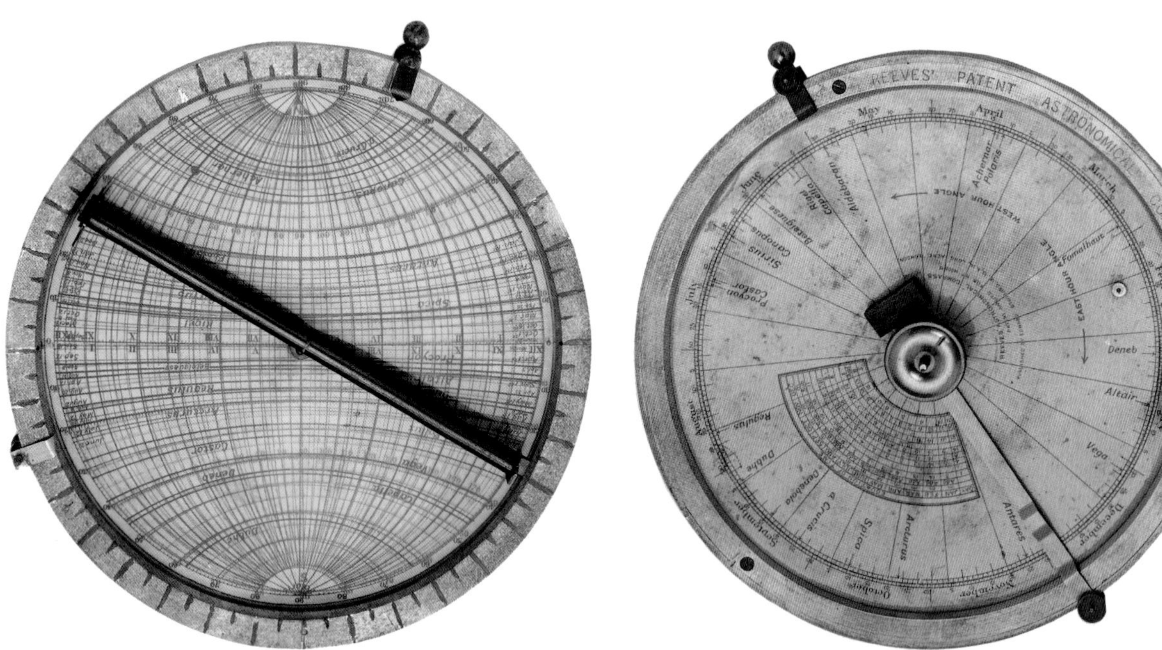

This modern la Hire astrolabe is used to determine both time and true north. The mater is an aluminum disc with two overlays, the inner one rotatable and made of cardboard and the outer one fixed and made of plastic printed in red. The rotatable disc has its limb divided by degrees, marked every five degrees and labeled every ten, 90°-0°-90°, twice from the top. The parallels, which show declination, are drawn every ten degrees. Heavier lines indicate the declination for specific stars. The 60°, 40°, and 20° parallels are divided into five-minute intervals. A list of dates near the limb, on both sides, shows the sun's declination on specific dates.

The meridians are drawn every five degrees and labeled along the equator, XII to I above and I to XII below. There are 22 stars named (★).

Over the rotatable disc is mounted a fixed plastic disc with the meridians drawn and the parallels printed in red. The parallels are drawn every five degrees and labeled on the circumference every ten degrees, 90°-0°-90°, twice from the top. The circumference is further divided by degrees. The meridians are drawn every five degrees and labeled every ten degrees, 0° to 180°, along both 40° parallels, each of which is further divided by degrees. "N" (with an arrow), "E," "S," and "W" are appropriately located. There is a rotatable rule on the face with a clamping nut to assist in taking the readings.

The reverse carries a "ready reckoner"[1] that has two scales on the limb. The outer scale is a civil calendar, divided into 365 days and marked and labeled every fifth day, with each month space having the appropriate number of day spaces. The month names are in English. The inner scale is divided every five minutes and marked every fifteen

● RIGHT ASCENSIONS

Star name	*Date*
Polaris	April 12
Achernar	April 14
Aldebaran	May 29
Capella	June 7
Rigel	June 7
Betelguese	June 18
Canopus	June 26
Sirius	June 30
Castor	July 12
Procyon	July 14
Regulus	Aug. 22
Dubhe	Sept. 4
Denebola	Sept. 16
α Crucis	Sept. 25
Spica	Oct. 9
Arcturus	Oct. 22
Antares	Nov. 26
Vega	Dec. 29
Altair	Jan. 17
Deneb	Jan. 29
Fomalhaut	March 3

OPPOSITE, LEFT: *Face of No. 30*
OPPOSITE, RIGHT: *Reverse of No. 30*

minutes with a short line and every hour with a long line toward the center point. In the upper part is an equation of time table, so labeled, showing the number of minutes to be added or subtracted to convert to mean solar time. Within the lower part is the inscription, curving around the center hole: "REEVES' ASTRONOMICAL COMPASS PATENT APPLIED FOR NO. 14008.08 PUBLISHED BY EDWARD STANFORD, 12, 13, & 14 LONG ACRE, LONDON." Below this, on the left, is printed "EAST HOUR ANGLE," with an arrow curved to left, and on the right "WEST HOUR ANGLE," with an arrow curved to right. The vertical hour line runs from September 20 at the top to March 21 at the bottom, indicating the first point of Aries.

Radially, from the outside in, 21 stars are listed, each with an arrow pointing to a specific date, according to their Right Ascension (●). Pollux is not shown.

There is a brief explanation of this instrument, with drawings, in Reeves' patent application, which accompanied the instrument and is now on file at the Adler Planetarium.[2]

1 ARIES = 21 March, from the calendar scales.
PROVENANCE — A. P. gift, 1985.

1. Saunders (1984), 70-71.
2. "Abridgment Class Philosophical Instruments 14,008," Reeves, E. A. July 1 (Cognate Application No. 14,523, A.D. 1908, dated July 31), p. 394.

★ STAR LIST

Star name	*Modern name*
Polaris	Alpha Ursae Minoris
Dubhe	Alpha Ursae Majoris
Capella	Alpha Aurigae
Deneb	Alpha Cygni
Vega	Alpha Lyrae
Castor	Alpha Geminorum
Pollux	Beta Geminorum
Arcturus	Alpha Bootis
Aldebaran	Alpha Tauri
Denebola	Beta Leonis
Regulus	Alpha Leonis
Altair	Alpha Aquilae
Betelguese	Alpha Orionis
Procyon	Alpha Canis Minoris
Rigel	Beta Orionis
Spica	Alpha Virginis
Sirius	Alpha Canis Majoris
Antares	Alpha Scorpii
Fomalhaut	Alpha Piscis Australis
Canopus	Alpha Carinae
Achernar	Alpha Eridani
α Crucis	Alpha Crucis

31 Classic-type astrolabe

Rob Lucas-Dean
Ecton Brook, England
1977
"Made by ROB LUCAS-DEAN, Ecton Brook, 1977.
Reproduced by courtesy of The Trustees of
the National Maritime Museum, Greenwich, England.
NMM REF. A37 / NA37-9 / C"
inside the mater
22.4 x 16.0 x 2.8 cm.
Rete thickness — 0.8 cm.
Brass
W-97

★ STAR LIST

Star name	Modern name
DELPHIN	Alpha Delphini
AQVILA	Alpha Aquilae
CAP. SERP:	Gamma Draconis
PAL:SI: SERP:	? Draconis
HVME: EQVI	Beta Pegasi
PEGASVS	Alpha Pegasi
CORONA	Alpha Coronae Borealis
ARCTVRVS	Alpha Bootis
CAVDA CETI	Beta Ceti
ANDROMEDA	Beta Andromedae
GORGON	Beta Persei
HOLOR	Alpha Cygni
VEGA	Alpha Lyrae
CAVDA LEONIS	Beta Leonis
CORVVS	Alpha Corvi
SPICA VIRGINIS	Alpha Virginis
VENTER CETI	Zeta Ceti
NARIS CETI	Alpha Ceti
SE'PES ORIONIS	Beta Orionis
OCVLVS ♉	Alpha Tauri
CAPRA	Alpha Capricorni
CANIS MAIOR	Alpha Canis Majoris
CANIS MINOR	Alpha Canis Minoris
HIDRA	Alpha Hydrae
COR LEONIS	Alpha Leonis
PATERA	Alpha Crateris

LEFT: *Face of No. 31*
RIGHT: *Reverse of No. 31*

This astrolabe is an electrotype copy of a typical Georg Hartmann astrolabe in the National Maritime Museum collections at Greenwich. It is inscribed on the back of the mater, below the shadow square: "GEORGIVS HARTMAN / NOREMBERGAE FACIEBAT / ANNO MDXLVIII."

The rete contains 26 stars (★).

PROVENANCE — National Maritime Museum, Greenwich; R. S. Webster, Winnetka, Ill., 1980; A. P. gift, 1980.

32 Azarquiel-type astrolabe

Roser Puig Aguilar
Barcelona
1985
"Azafea / ŠAKKĀZĪYA / de Azarquiel /
Tesis doctoral de Roser Puig Aguilar / Barcelona, 1985"
above the shadow square
18.5 x 15.2 x 1.5 cm.
Rete thickness — 0.5 cm.
Cardboard, paper, and plastic
A-251

★ STAR LIST

Star name	*Modern name*
Dubhe	Alpha Ursae Majoris
Alkaid	Eta Ursae Majoris
Altair	Alpha Aquilae
Aldebaren	Alpha Tauri
Sirius	Alpha Canis Majoris

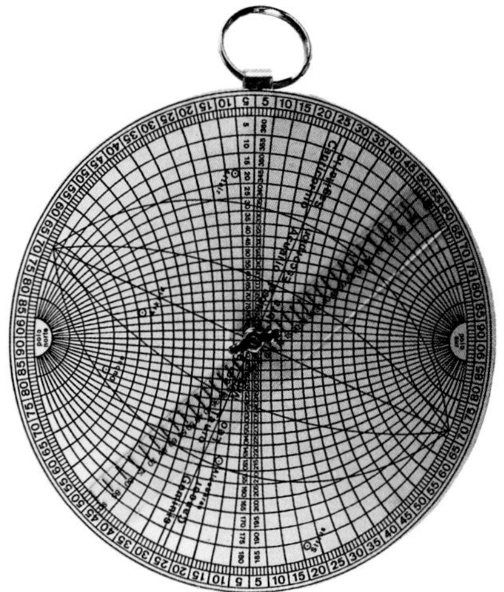

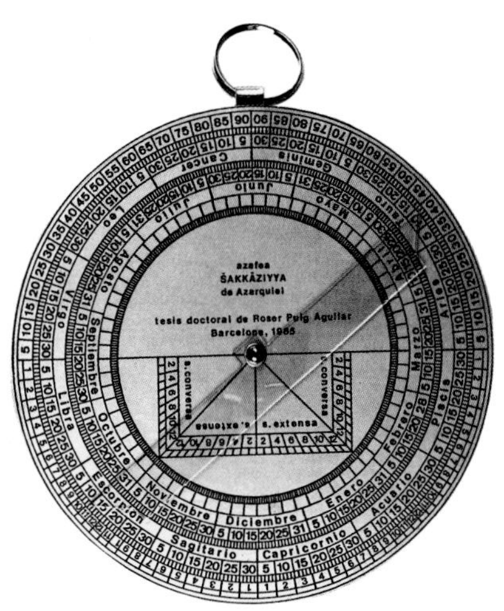

LEFT: *Face of No. 32*
RIGHT: *Reverse of No. 32*

The face carries a circle of degrees and an Azarquiel-type projection, with the North Pole on the left (★). The limb is divided every degree and labeled every five degrees, 0°-90°-0°, twice from the top. The parallels are drawn every five degrees, as are the meridians. The latter are labeled above the center line from the right, 5° to 180° (upside down), and below the line from the left, 185° to 360°. The line of the ecliptic is divided into six parts, each labeled with the names of two zodiacal signs in Spanish.

The poles are 90° from the top ring and are labeled "polo sur" and "polo norte." The plastic regula is divided and labeled every five degrees, 0° to 90°, twice from the center.

The reverse carries nine scales, including a scale of degrees, labeled 90° to 0° twice from the ring to the mid-line, and a scale in the lower half, labeled 1-12-12-1 twice clockwise. Then come the calendars, zodiacal and civil, appropriately divided and labeled in Spanish. Lastly, there is another circle of degrees divided every five degrees and then every degree.

The inscriptions are above the shadow square, which is divided into twelve parts and labeled 2 to 12 four times. "s. conversa" and "s. extensa" are each labeled twice.

PROVENANCE — Roser Puig Aguilar, Barcelona, 1985; A. P. gift, 1985.
REFERENCE — Puig (1986).

33 Classic-type astrolabe

Norman Greene
Berkeley, California
1989
Unsigned
20.3 x 15.5 x 1.9 cm.
Rete thickness — 0.7 cm.
Brass
W-250

This modern astrolabe is loosely copied from Geoffrey Chaucer's[1] and the Painswick[2] astrolabes.[3]

The brass mater is cast, with the cavity turned out on a lathe. The throne is a knob supported by a step on either side. The bail and ring pivot. The limb on the face carries the alphabet (missing I, U, and W), with the letters separated by diamond-shaped devices. A degree scale is marked every five degrees. The cavity is blank.

On the reverse, the limb shows two concentric calendar scales. The shadow square is divided into twelve parts and marked and labeled by threes.

The rete has 20 star points (★).

The tympan is for 34°, and the unequal hours are shown.

There is an alidade. The rule is double and unmarked. The bolt is equipped with a nut instead of a horse.

I ARIES = 21 March, from the calendar scales.

PROVENANCE — Norman Greene, Berkeley, Calif., 1989; R. S. Webster, Winnetka, Ill., 1989; A. P. gift, 1989.

★ STAR LIST

Star name	*Modern name*
Mark	Alpha Pegasi
Deneb	Alpha Cygni
Altar	Alpha Aquilae
Vega	Alpha Lyrae
Rosalhague	Alpha Ophiuchi
Alpheca	Alpha Coronae Borealis
[unnamed]	Alpha Scorpii
[unnamed]	Alpha Virginis
Arcturus	Alpha Bootis
Al	Alpha Ursae Minoris
B	Beta Ursae Minoris
Alfard	Alpha Hydrae
Procyon	Alpha Canis Minoris
Sirius	Alpha Canis Majoris
Betelgeuse	Alpha Orionis
Rigel	Beta Orionis
Aldebora	Alpha Tauri
Menkar	Alpha Ceti
Diphda	Beta Ceti
Alfer	Beta Pegasi

1. Chaucer (1391); Chaucer (1872).
2. Gunther (1932), 475, plate 131.
3. Greene (1977).

34 Classic-type astrolabe

Norman Greene
Berkeley, California
1989
Unsigned
21.0 x 15.5 x 1.9 cm.
Rete thickness — 0.7 cm.
Brass
W-251

This astrolabe is identical to W-250 (cat. no. 33) except for slight differences in overall height and in tympan latitude. Greene astrolabes usually come with a 34° tympan, but the W-251 tympan, made as a special order, is for 42°.

I ARIES = 21 March, from the calendar scales.

PROVENANCE — Norman Greene, Berkeley, Calif., 1989; R. S. Webster, Winnetka, Ill., 1989; A. P. gift, 1989.

35 Classic-type astrolabe

England
c. 1989
"A.RT." *on the rete*
6.4 x 5.1 x 0.6 cm.
Rete thickness — 0.3 cm.
Brass
W-253

★ STAR LIST

Star name	*Modern name*
Diphda	Beta Ceti
Deneb	Alpha Cygni
Markab	Alpha Pegasi
Altair	Alpha Aquilae
Vega	Alpha Lyrae
Antares	Alpha Scorpii
Arcturus	Alpha Bootis
Alkaid	Eta Ursae Majoris
Spica	Alpha Virginis
Regulus	Alpha Leonis
Alphard	Alpha Hydrae
Procyon	Alpha Canis Minoris
Betelguese	Alpha Orionis
Capella	Alpha Aurigae
Sirius	Alpha Canis Majoris
Rigel	Beta Orionis
Aldebaren	Alpha Tauri
Menkar	Alpha Ceti

LEFT: *Face of No. 34*
RIGHT: *Reverse of No. 34*

Face of No. 35

This miniature brass astrolabe is more a piece of jewelry than a scientific instrument. The throne is small and carries a ring. The limb on the face has an hour scale labeled XII to XII twice. The inner scale is divided into 360 degrees and labeled 90°-0°-90° from the top. The reverse is blank. The cavity has a tympan engraved in it.

The rete vaguely resembles Chaucer's and carries eighteen stars (★). There are no real star points, though some are indicated by arrows. Some are wrongly placed. "A.RT." is stamped between Sirius and Rigel, opposite the index point.

The tympan, which is for 49°, is engraved inside the mater. The azimuths and almucantars are drawn and labeled every ten degrees. The rule is double and is divided every ten degrees. One side is labeled 40°, 20°, 0°, 20°. The other reads 30°-0°-20°, with 0° being on the equatorial line, in order to determine the declination. The rule is riveted on. There is no alidade.

PROVENANCE — National Maritime Museum Book Shop, Greenwich, *c.* 1989; R. S. Webster, Winnetka, Ill., 1991; A. P. gift, 1991.

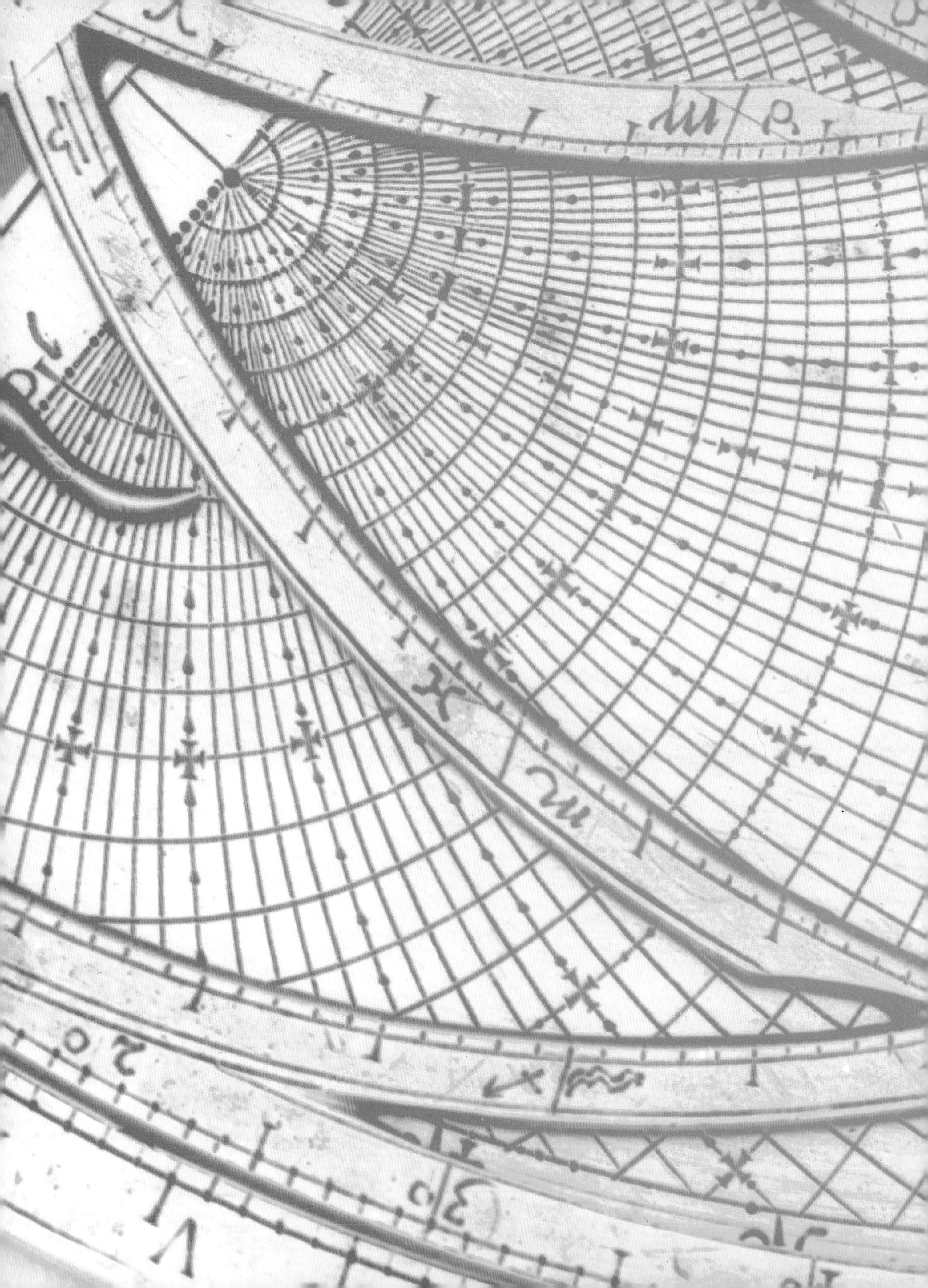

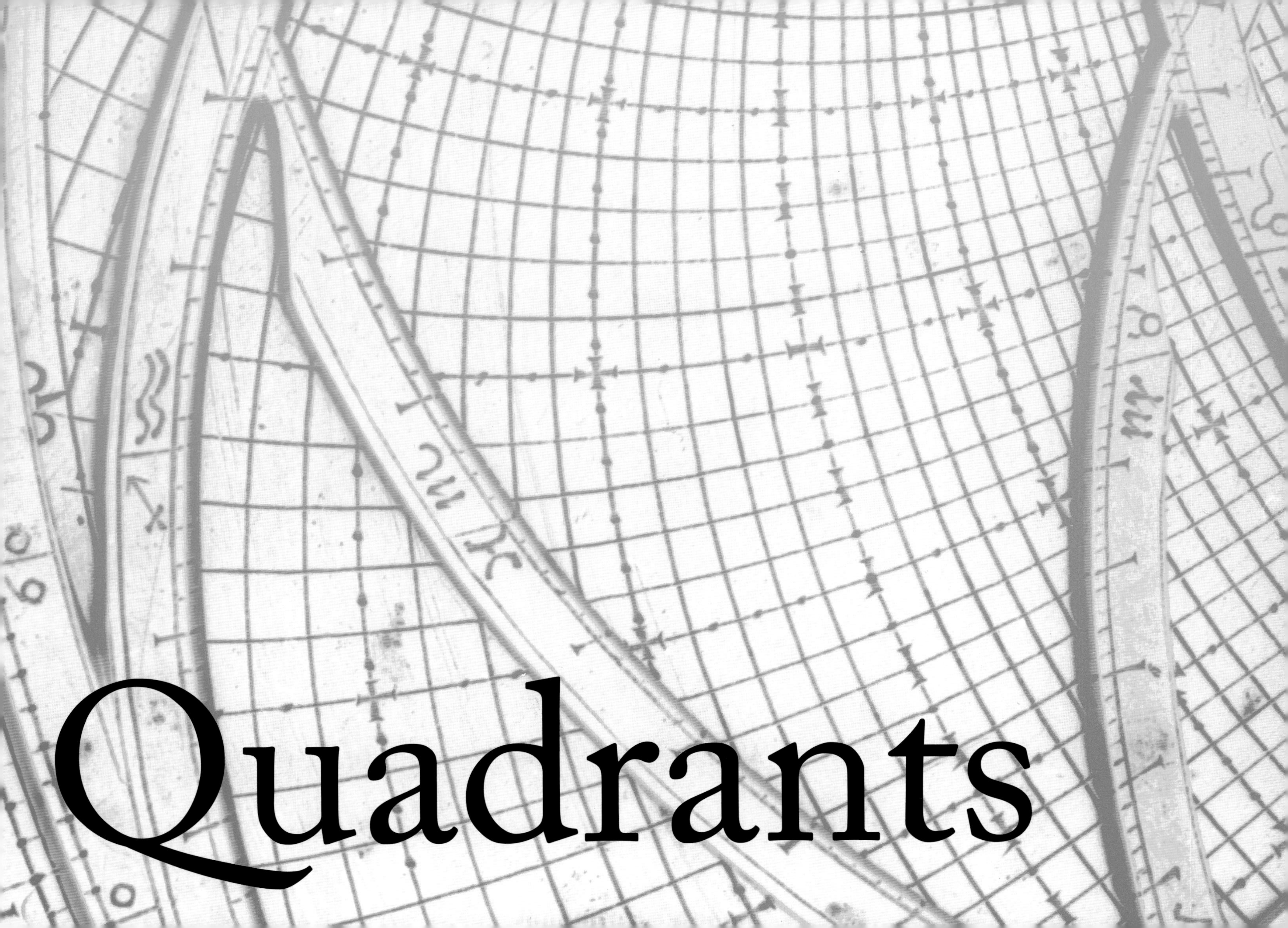

Quadrants

FIGURE 1 *Quadrants in Use. Lansberg (1635), title page.*

The Astrolabe-Quadrant

Another instrument based on the planispheric projection is the astrolabe-quadrant. First described about 1260 by the Jewish astronomer and mathematician Jacob ben Machir ibn Tibbon (*c.* 1236-1305), it became known as the new quadrant of Prophatius, after his Latin name. (Figure 2)

A very popular form of the astrolabe-quadrant was developed by the Englishman Edmund Gunter (1581-1626) about 1620. (Figure 3) The stereographic projection is from both the North and the South Celestial Poles onto the plane of the equator.

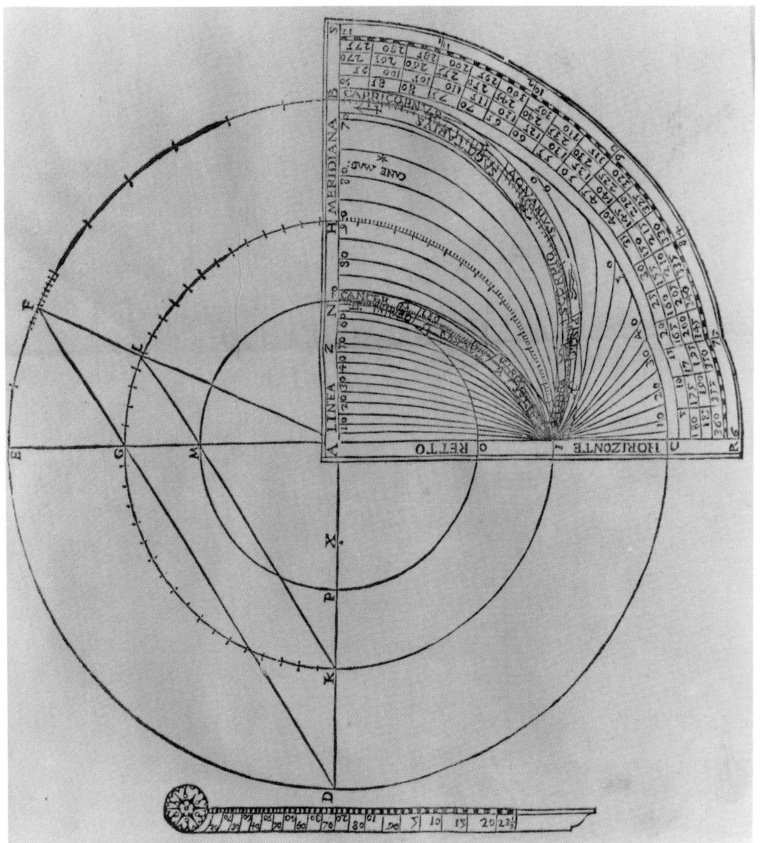

FIGURE 2 *Profatius Quadrant. Gallucci (1597), between fols. 76 and 77.*

As with most quadrants, it is engraved on a flat plate of wood or metal having the shape of a quarter circle. The rounded edge, called the limb, is divided by degrees from 0° to 90°. Inside the degree scale there is a calendar scale drawn so that on any date, the noon altitude of the sun may be read on the scale of degrees.

Inside the calendar scale, using the apex of the quadrant as the center, an arc is drawn that represents both the Tropics of Cancer and Capricorn. Within this arc, using the same center, another arc is drawn representing the celestial equator. This type of astrolabe-quadrant is figured and drawn for a specific latitude. On the left radius, from the equator to the Tropics is a scale of degrees from 0° to 23½° marking the declination. The right radius carries two pinhole sights for making altitude observations.

To complete the instrument, a thread with a plumb-bob is fastened at the apex of the quadrant. A bead or pearl is free to slide on the thread to aid in measuring the declination.

The right half of the projection carries the azimuth lines, curving to the left for summer and to the right for winter and marked along the equator, 10°, 30°, 60°, 90° unequally, and along the right radius from the equator down, 90° to 120°. In the left half are the hour lines, curving to the left for winter and to the right for summer, labeled along the equator, 6 to 12, and along the Tropics, 8 to 1, from left to right in both

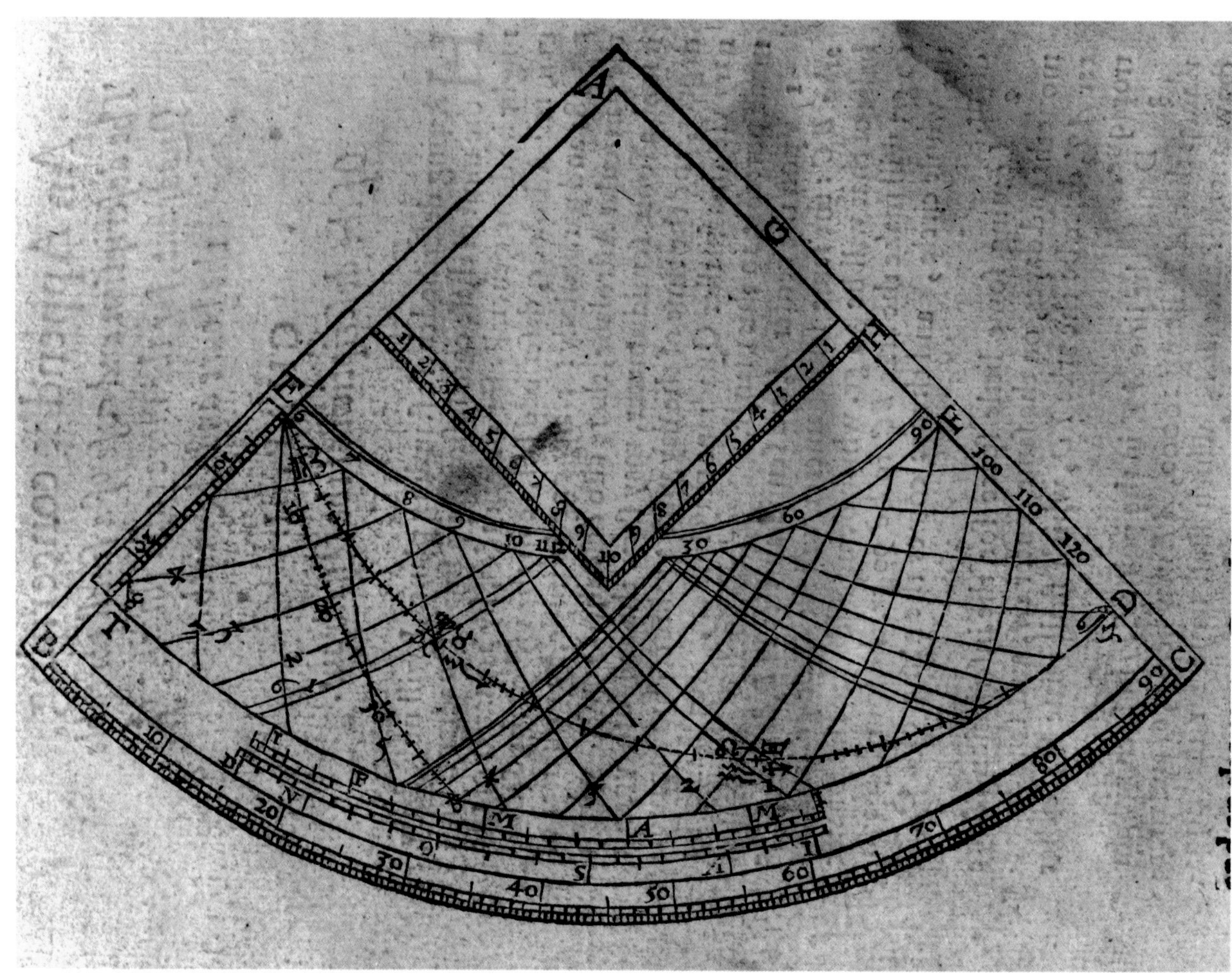

FIGURE 3 *Gunter's Quadrant. Gunter (1624), 188.*

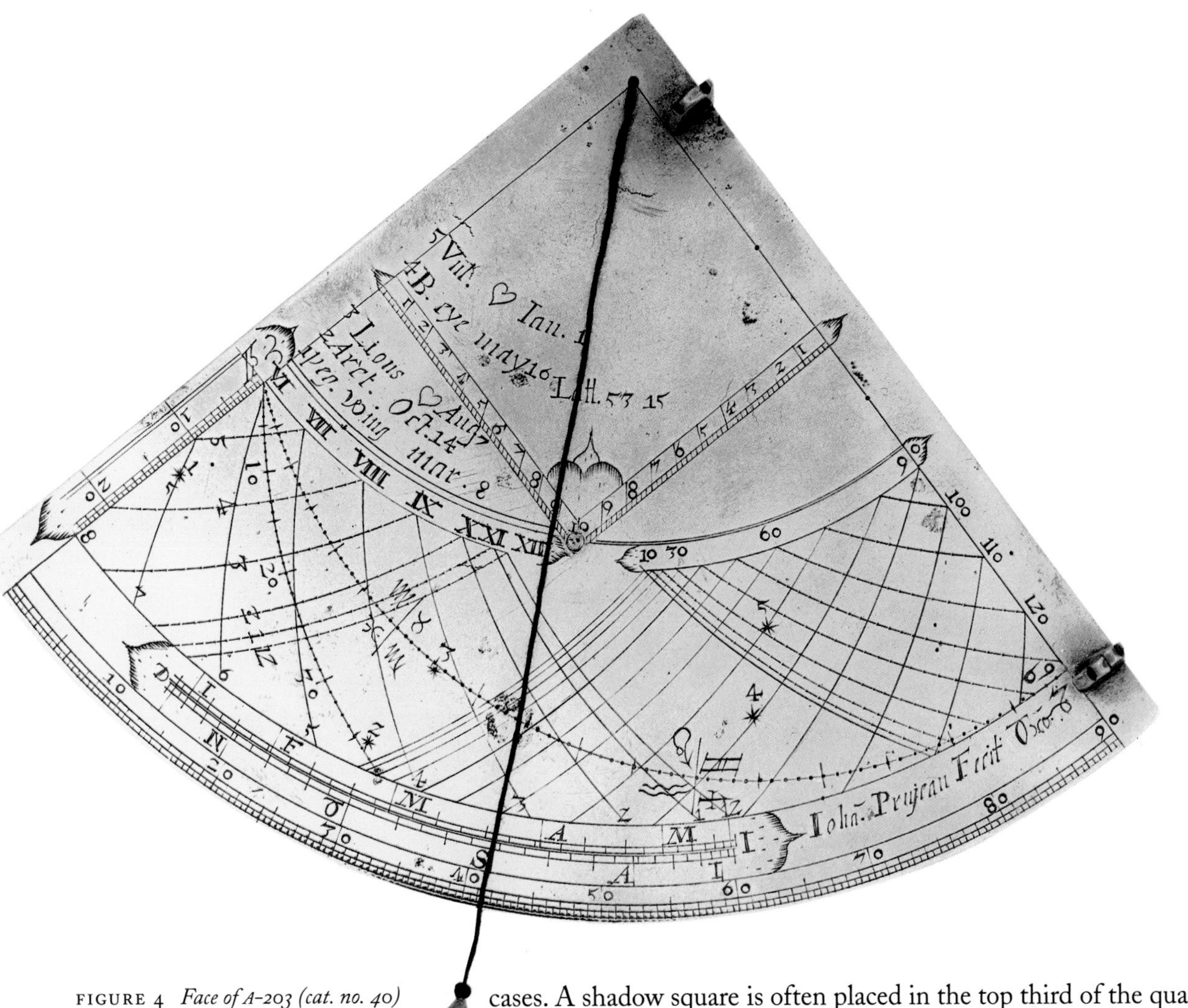

FIGURE 4 *Face of A-203 (cat. no. 40)*

cases. A shadow square is often placed in the top third of the quadrant above the line of the equator. (Figure 4)

In addition to the Tropics and the equator, the stereographic projection shows the ecliptic circle, horizon, hour lines, and azimuths. There are also a number of stars plotted to show their Right Ascension and declination. Generally the same five stars are listed above the shadow square with the date when they cross the meridian at midnight. The following example, from A-203 (cat. no. 40), is representative:

5	Vul. ♡	Ian. 1	[Vulture's Heart]	[Alpha Aquilae]
4	B. eye	May 16	[Bull's Eye]	[Alpha Tauri]
3	Lions ♡	Aug. 7	[Lion's Heart]	[Alpha Leonis]
2	Arct.	Oct. 14	[Arcturus]	[Alpha Bootis]
1	Peg. Wing	Mar. 8	[Pegasus' Wing]	[Alpha Pegasi]

Thus, with a Gunter-type quadrant it is possible to find the time and the direction from the height of the sun or from any of the stars shown on the instrument.

Other types of astrolabe-quadrants are discussed in the individual entries that follow.

36 Universal astrolabe-quadrant

Louvain
c. 1550
Unsigned
8.7 x 8.7 x 0.3 cm.
Working radius — 8.0 cm.
Brass
ICA 2079
A-108

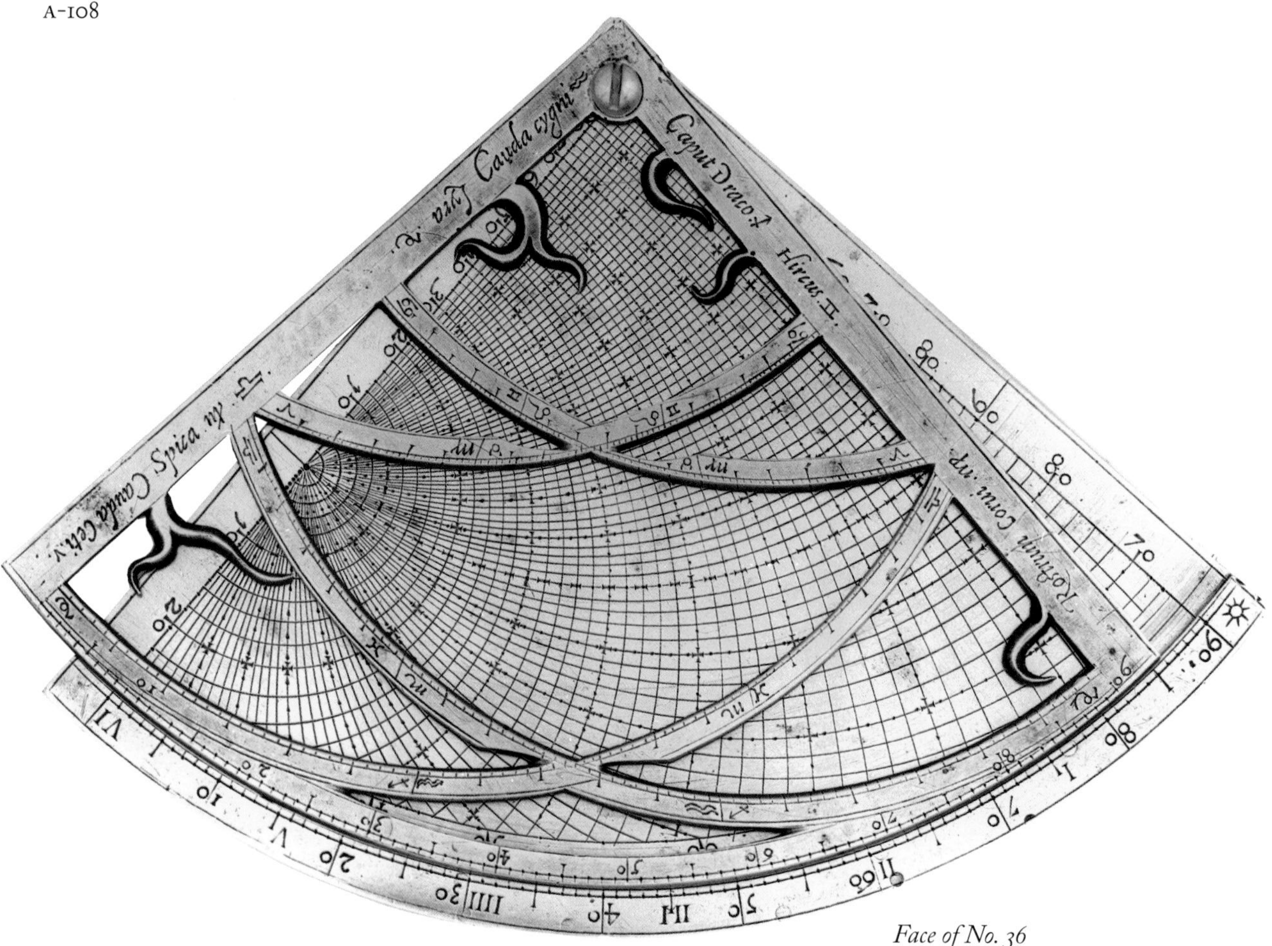

Face of No. 36

We have chosen to call this remarkable instrument a "universal astrolabe-quadrant" because it has a rete that is folded twice to form a quadrant, showing seven stars. The back has a reversible grid similar to Apianus' Meteoroscopion.[1] David A. King of the University of Frankfurt suggests that the provenance is Louvain, *c.* 1550, and that it be called a "universal horizon/azimuth quadrant."

The face of the quadrant carries a projection for 0° latitude. The limb is divided by degrees, marked every five and ten degrees, and labeled 10° to 90° counter-clockwise. The same scale carries an hour scale divided every four minutes, marked every 20 and 60 minutes, and labeled I to VI clockwise. There is a solar symbol ☉ at the right-hand end. The left side of the projection carries the declination (almucantar) scale, divided every two degrees, 23.5°-0°-90°, and labeled 20°-0°-80°. The azimuths also originate from the zenith and are divided every two degrees. They are labeled 10° to 60° along the inner limb and continue on the right edge, 70°-90°-10° toward the apex. Both almucantars and azimuth lines are ticked every ten degrees.

The rete carries the ecliptic circle twice, in four arcs, the Tropic of Capricorn, and seven stars, each with its zodiacal sign. The rete is first folded through the first points of Capricorn and Cancer and then, secondly, through the first points of Aries and Libra. Then the mirror image is superimposed. The back is plain.

David A. King has also studied this instrument. His comments are as follows:

"The rete which rotates over this side of the instrument is in the form of a quadrant. The ecliptic is represented twice, with the equinoxes on the meridian axis and on the perpendicular axis. In each case the two significant quadrants of the ecliptic

★ STAR LIST

Star name	*Zodiac symbol*	*Modern name*
Cauda Ceti	♈	Beta Ceti
Spica ♍	♎	Alpha Virginis
Cauda cygni	♒	Alpha Cygni
Lyra	♑	Alpha Lyrae
Caput Draco	♐	Alpha (Lambda) Draconis
Hircus	♊	Alpha Aurigae
Rostrum Corui	♍	Alpha Corvi

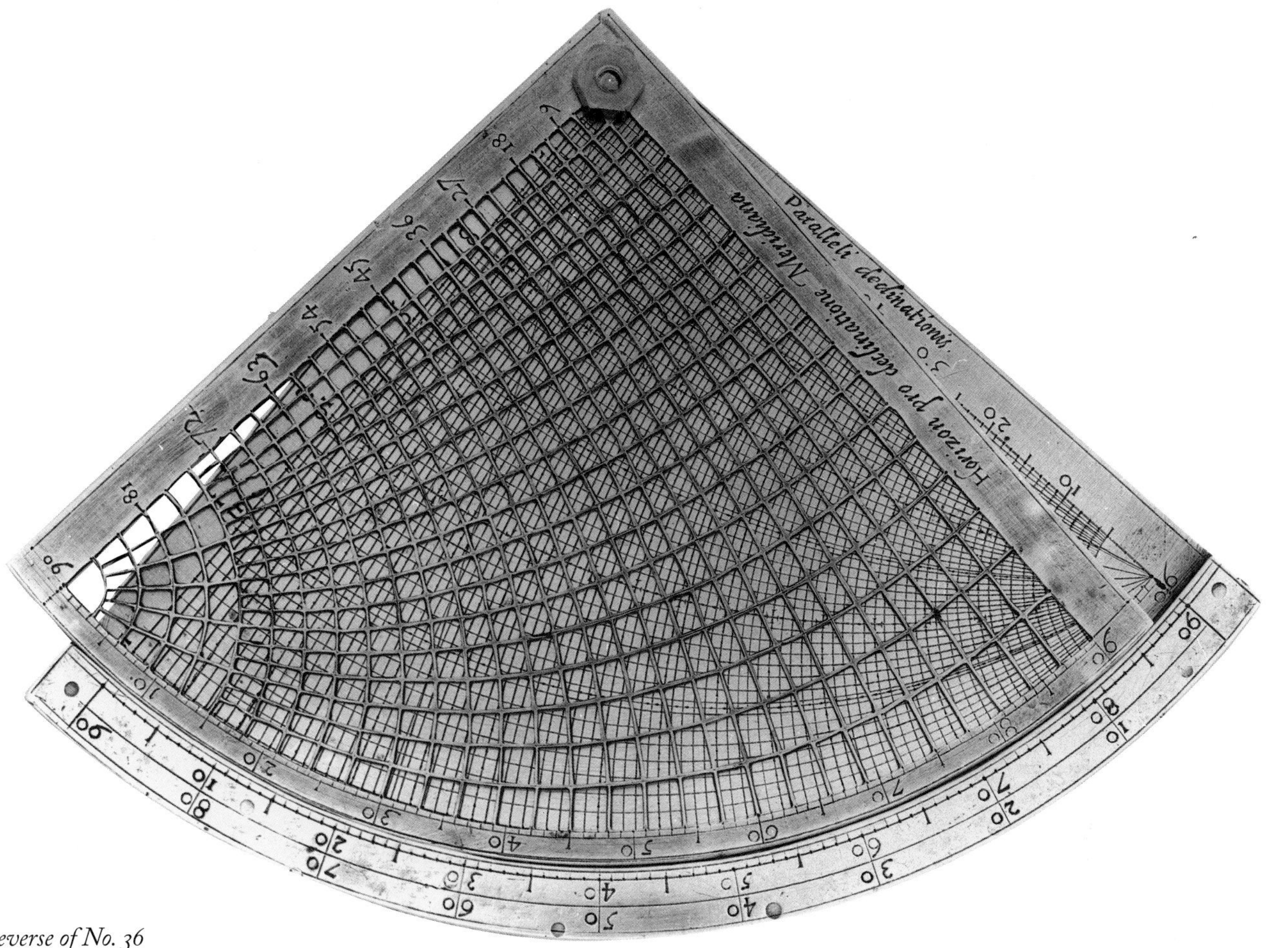

Reverse of No. 36

have been superimposed in such a way that the declination of any point on the ecliptic can be read from the meridian declination scale when that point lies on the meridian. The scale of the ecliptic is divided 5°/1° for each pair of signs, which are identified by their symbols. There is a scale around the rim which is divided 10°/5°/1° and labeled for each 10°, as on the quadrant underneath. There are pointers for seven stars, identified as follows (★), with symbols for the appropriate zodiacal signs so that the user knows to which quadrant of the ecliptic they correspond in longitude.

"The pointers are of the flame-shaped variety common on medieval Catalan Italian and especially French astrolabes as well as Renaissance Flemish ones; in the modern literature these are associated with the instruments of Jean Fusoris (Paris, *c.* 1365-1436) but their origin is earlier."

The reverse also carries, as King comments, "a set of astrolabic markings for latitude 0° bounded by the circle corresponding to the celestial equator. [The limb carries two scales, both for 0° to 90°. The outer one is labeled 10° to 90° and the inner one 90° to 10°, both clockwise.] There are altitude circles [arcs] for each one degree and azimuth circles [arcs] for each two degrees. The former are fish-boned [or ticked] for each five degrees and the latter for each fifteen degrees (which means that three additional curves are engraved). The altitude arguments [or lines] are labeled along the vertical [meridian] axis [right edge] for each ten degrees as declinations and around the circumference as altitudes. The azimuth markings for each fifteen degrees are labeled for the hours 7/5-8/4-11/1-12 on the [left-hand] horizontal axis from the center." The scale on the outer limb shows declinations, while the inner shows altitudes. A line, from the apex to 23.5° on the limb, represents the

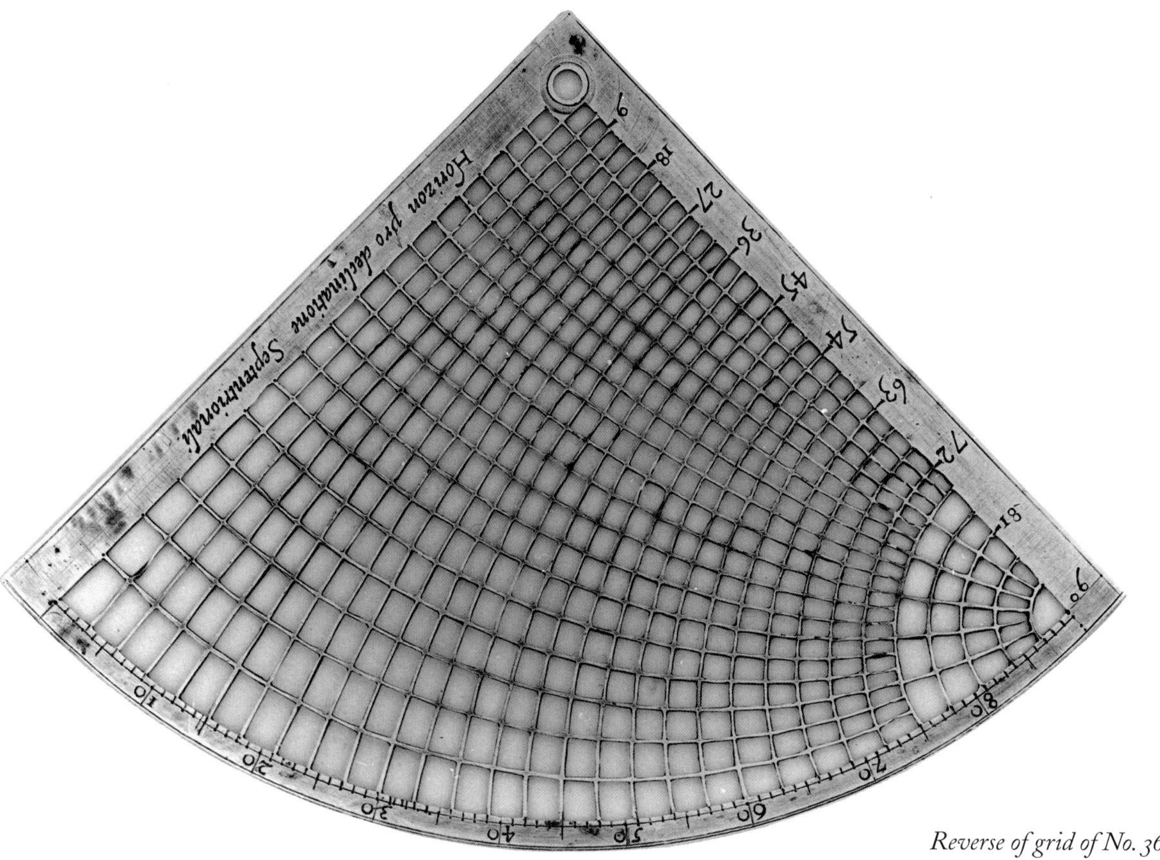

Reverse of grid of No. 36

ecliptic. The right edge of the projection is labeled "Parallelli declinationis."

Continuing, King states that the grid, "which can move over this side of the instrument, is engraved on both sides and must be turned over according to need. One side is marked 'Horizon pro declinatione Meridiana' and the other 'Horizon pro declinatione Septentrionali.' It consists of a grid of šhakkāzīya curves for each 3° of altitude and each 5° of azimuth. The altitudes are labeled for each 9° on the vertical axis. The outer scale is divided 10°/5°/1° and labeled for each 10° of altitude on the 'Septentrionali' side and each 10° of declination on the 'Meridiana' side. The azimuth circles are for each 15° above altitude 75° and there are none above altitude 87°."

Koenraad Van Cleempoel, a Ph.D. candidate at the Warburg Institute, believes that there is a similarity between the engraving on A-108 and the re-engraving of the ecliptic circle on the rete and of the plate for latitude 40° 30' of the "Philip II" astrolabe at the Madrid Archaeological Museum.[2]

This instrument came on a much later stand. The bolt and nut are modern replacements.

PROVENANCE — Found in 1941 in a Mexico City antique shop by the son of Leo Metzenberg, a friend of Max Adler; A. P. purchase, 1941.

REFERENCES — Gibbs *et al.* (1973), 88; King (1974), plate 1; King (1991), plate on 175.

1. Apianus (1540), fol. M-4.
2. For a fuller discussion, see Van Cleempoel (1997).

37 Astrolabe-quadrant

Gunter-type
Anthony Thompson
London
c. 1650
"Anthony Thompson *in Hosier lane neer Smithfield"*
along the right radius
11.8 x 11.7 x 0.1 cm.
Working radius — 11.0 cm.
Brass
DPW-10

★ STAR LIST

* p n	[Pegasus' Wing]
* a	[Arcturus]
* l h	[Lion's Heart]
* b e	[Bull's Eye]
* v h	[Vulture's Heart]

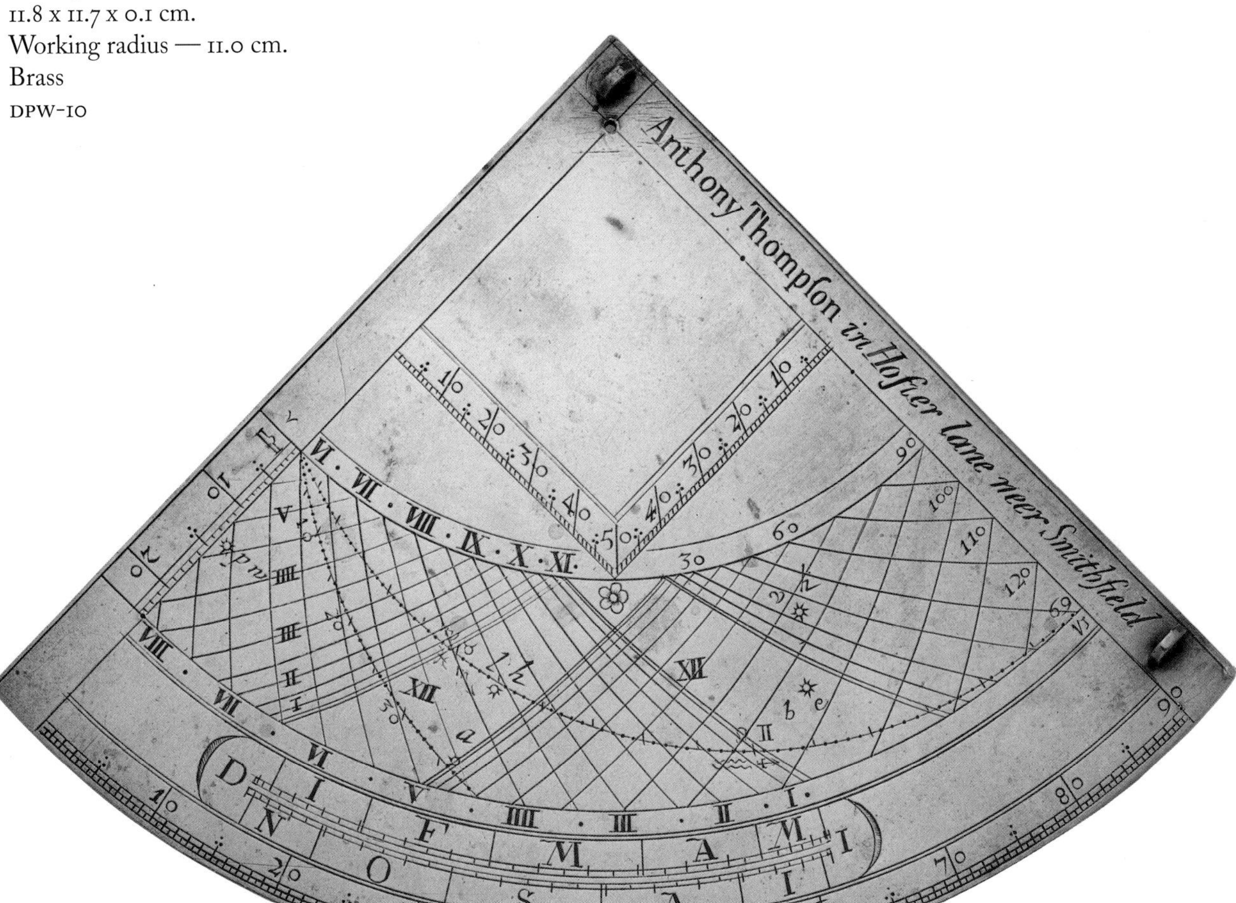

Face of No. 37

The face carries a Gunter-type projection laid out for the latitude of London. The winter hours are labeled VI to XI at the top and V to I (in reverse) below. The summer hours are labeled VIII to I (in reverse). The azimuths are labeled 30°, 60°, 90°, 100°, 110°, and 120°. The shadow square is divided every part, marked every five parts, and labeled every ten parts, 10-50-10. The usual five stars are listed (★). Pegasus' Wing is mislabeled "p n."

The calendar is labeled with the months' initials, with "J" shown as "I." The months are divided as usual. The signature appears along the right radius: "Anthony Thompson *in Hosier lane neer Smithfield."*

The sights are present, but the plumb-bob is not original. The reverse is plain.

PROVENANCE — David P. Wheatland, Topsfield, Mass., 1986, no. 694 = no. 359; A. P. gift, 1986.

38 Astrolabe-quadrant

Sutton-type
Henry Sutton
London
1658
"Latit / ·51·32· / 1658 / Henr Sutton Londini / fecit."
on the face
13.8 x 13.5 x 0.01 cm.
Working radius — 11.7 cm.
Paper, printed from
an engraved plate
W-256

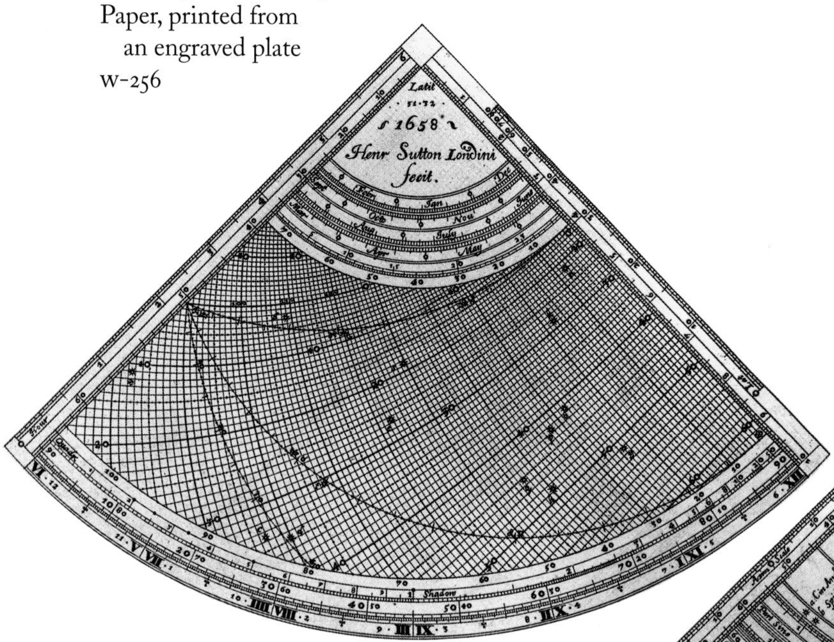

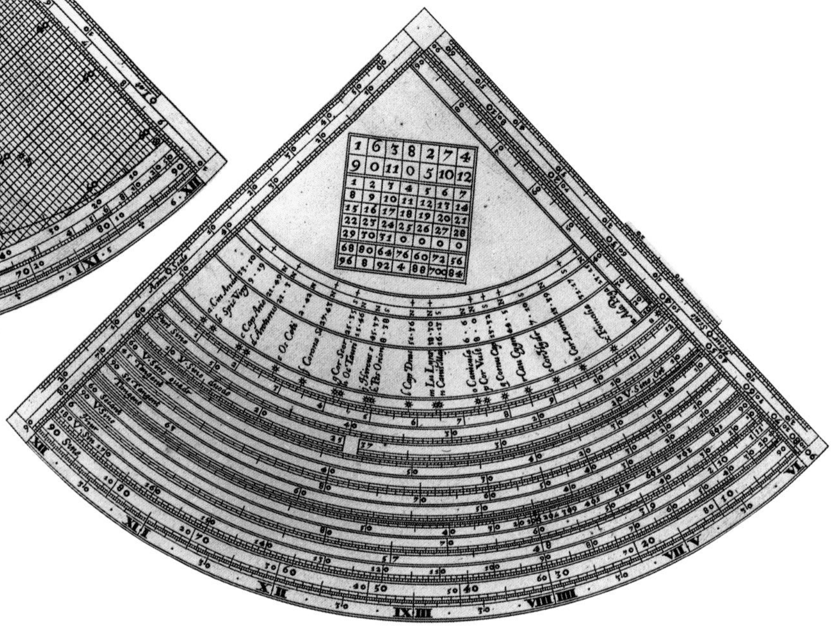

LEFT: *Face of No. 38*
RIGHT: *Reverse of No. 38*

This quadrant was designed by Henry Sutton, and its construction and multiple uses were described by John Collins in his book, *The Sector on a Quadrant,* in 1658. This example at the Adler is, in fact, the frontispiece from a copy of Collins' work and carries a note to the binder, "Place this next after the Title page." It was found loose between the pages of an example of Edmund Stone's translation of Nicolas Bion[1] that was given to the Adler. The engraving is slightly under the size described by Collins.

The quadrant is an inverted Stöffler projection, with the Tropic of Cancer near the limb and the Tropic of Capricorn forming the upper boundary. It is bordered by a line of equal parts on the right edge and a line of tangents on the left. Above the quadrant projection the azimuths are labeled 10° to 70° clockwise. Then comes the line of declination, labeled 10° to 23° counter-clockwise, followed by the calendar scale set out in four arcs.

Below the quadrant the azimuths are also labeled and divided every ten degrees. Next, in the left half, is the Geometrical Quadrant, labeled "Quadr" and 1 to 1[0]. The right half contains the line of shadows, labeled "Shadow" and 1 to 6, 8, 10, 20, and 50. The limb carries the quadrant scale, divided every quarter of a degree and labeled 10° to 90° and 90° to 10°. At the outer edge are four hour scales, divided into minutes and labeled XII to VI and 6 to 12 clockwise and VI to XII and 12 to 6 counter-clockwise.

On the right wing, or edge, there is a latitude scale, labeled "Lat" and 0° to 90° from the limb to the apex. The left wing carries a scale of hours, labeled "Hour" and 0 to 6 from the limb.

The upper part of the quadrant carries the inscription: "Latit / ·51·32· / 1658 / Henr Sutton Londini / fecit."

Fourteen stars are indicated on the quadrant by asterisks and letters of the alphabet. These are taken from the list of 21 stars radially listed on the reverse.

The lines for the summer and winter ecliptic arcs are shown with the appropriate signs. The horizon line is also drawn. A plumb-line and bead are necessary in order to do the calculations and use the instrument.

The reverse carries various scales, a radial list of 21 stars (★), and an almanac. The quadrant is bordered by a line of sines on the right side, labeled from the apex 10° to 90° and also 90° to 10° in smaller figures. On the left side is a line of chords, labeled 10° to 90° from the apex. On the limb, the hour scale runs both directions XII to VI, with the half hours labeled 30. Next is the quadrant scale, divided by degrees and labeled 0° to 90° and 90° to 0°. It is mislabeled "Sine." Following are thirteen scales often found on a sector. They are, from the limb inward to the apex, running clockwise unless noted:

1) Sine scale.

2) "V. Sin" (Versed Sine), 0° to 180°.

3) "Hour" (Hour scale), 1 to 6 and 6 to 11 in reverse, with uneven spacing.

4) "V. Sine" (Versed Sine), 0° to 90°.

5) "Secant" (on the left side of the arc), 0° to 60°, starting at the VIII-IIII point on the limb.

6) The lesser sine (on the right side of the arc of the fifth scale), from the edge to the VIII-IIII point and reverse, 0° to 90° and 90° to 0° (on the same arc).

7) "Tangent," 0° to 63° 26'.

8) "Se Tangent" (Semi-Tangent), 0° to 90°.

9) "Tangent," 0° to 45°.

10) "V. Sine quadr" (Versed Sine quadrupled [to four times the radius]), 0° to 60°.

11) "V. Sine duode" (Versed Sine duodece [to twelve times the radius]) on the left, 25° to 0°.

12) "V. Sine Oct" (Octupled Versed Sine [to eight times the radius]) (on the right side of the arc of the eleventh scale), 0° to 37°.

13) "Part Sine" (Particular Sine scale), for the latitude 51° 32'.

★ STAR LIST

Star name	*Declination*	*Modern name*
a. Cin Andr	33° 50' N	Beta Andromedae
b. Spic Virg *	9° 19'	Alpha Virginis
c. Cap Arie *	21° 49' N	Alpha Arietis
d. Arcturus *	21° 3' N	Alpha Bootis
e. Os Ceti *	2° 42' N	Alpha Ceti
f. Corona Se	27° 32' N	Alpha Coronae Borealis
g. Cor Scor *	25° 35' S	Alpha Scorpii
h. Oc Tauri *	15° 46' N	Alpha Tauri
j. Hircus s	45° 37' N	Alpha Aurigae
k. Pes Orion *	8° 38' S	Beta Orionis
l. Cap Drac	52° 36' N	Beta Draconis
m. Lu Lyrae	50° 30' N	Alpha Lyrae
n. Canis Maj *	16° 13' S	Alpha Canis Majoris
o. Canicula *	6° 6' N	Alpha Canis Minoris
p. Cor Vult *	8° 0' N	Alpha Aquilae
q. Cornu Cap *	15° 52' S	Alpha Capricorni
r. Cau Cygn	44° 5' N	Alpha Cygni
s. Cor Hydr *	7° 10' S	Alpha Hydrae
t. Cor Leoni *	15° 59' N	Alpha Leonis
x. Fomaha	31° 17' S	Alpha Piscis Australis
z. Ala Pega *	13° 25' N	Gamma Pegasi

"The Nocturnal with stars," as Collins called it, comes next. The Right Ascension is given in hours, 1 to 12 counter-clockwise, divided every five minutes. The radial list of 21 lettered stars is each preceded by a star symbol and a letter (★). They are all followed by their declination, most of which are marked "N" or "S." The last arc contains eleven "+" marks. The meaning of these marks is unclear, as they are not explained by Collins, but they do not relate to magnitude, declination, or the position of the star on the face of the quadrant.

An additional, shorter line of sines lies along the inner right edge of the quadrant, from the outer edge of the star list to the apex, labeled 10° to 90°. This scale has misplaced the name "Ala Pega." However, its star symbol and a small "z" indicate the correct Right Ascension for this star.

The "almanack," as Collins called it, occupies the upper part of the quadrant. The seven vertical columns represent the days of the week, starting with Sunday. The first two horizontal rows are for the months of the year, starting with March. The next five rows are the days of the month, and the last two rows are used during leap years.

On the right wing, the first scale, labeled "Hour Scale," runs from the limb inward, 23.5°-0°-62°. A condensed declination scale lies outside of this, from the 0° point toward the apex. The left wing carries a third scale, labeled "Azim Scale" and laid out the same way, 23.5°-0°-62°, on which the same condensed declination scale is repeated.

1 ARIES = 10 March, from the calendar scales.

PROVENANCE — R. S. Webster, Winnetka, Ill., 1970; A. P. gift, 1984.

1. Stone (1723).

39 Astrolabe-quadrant

Panorganon, or universal instrument
Walter Hayes
London
c. 1680
"W. Hayes fecit / Dublin 1740"
(latter added in a different hand)
on the face
15.3 x 15.3 x 0.1 cm.
Working radius — 13.9 cm.
Brass
T-35

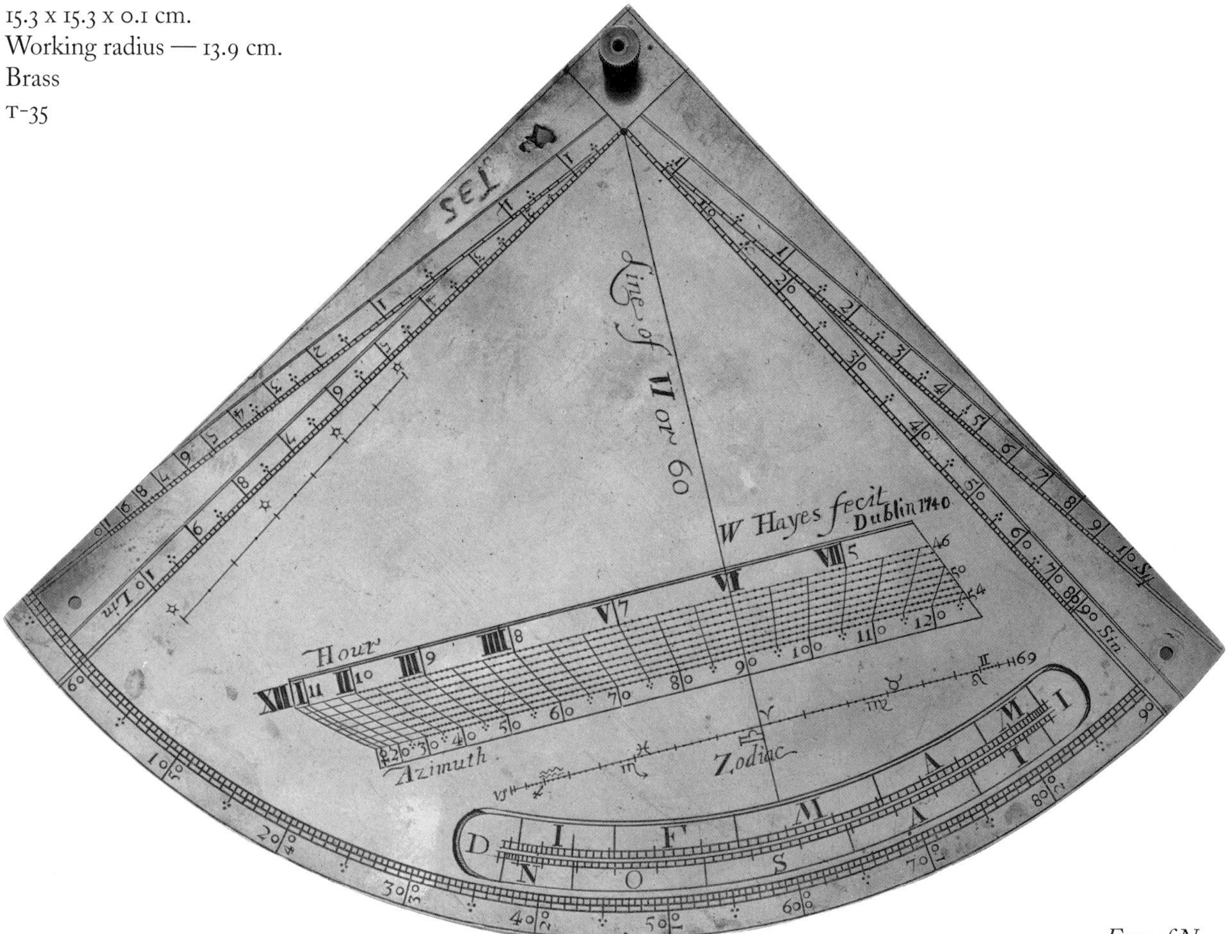

Face of No. 39

The Panorganon, or universal instrument, differs from a Gunter-type quadrant in that it is drawn for several latitudes, not just for a particular latitude. It was described by William Leybourn in 1672.[1] He gave credit, in the introduction, to Samuel Foster for many of the ideas developed.

Leybourn said that this quadrant was part of the "analemma," which in this case referred to an orthographic projection of the celestial sphere,[2] the same projection as is shown on the de Rojas-type astrolabe. In the quadrant only enough of the projection is shown to deal with problems concerning times and azimuths of the sun, stars, etc. The use of the word analemma first appeared in 1653.[3]

The scale on the limb is divided two ways. The first is from 0° to 90° counter-clockwise. The other is divided with 00° at the line of 60 and runs clockwise to 66.5° and counter-clockwise to 30°. This is the declination scale and extends into the left wing.[4] 65°, 55°, etc. are marked by three small dots in a triangle. The civil calendar is laid out above the declination scale between 23.5° on the left to 23.5° on the right. The months are indicated by initials, from the left "D," "I," "F," etc. above and "D," "N," "O," etc. below. It is divided into two-day intervals. Above the civil calendar lies the line of the "Zodiac," so labeled. The zodiacal signs are shown.

Above the zodiacal line, the hour azimuth diagram is drawn showing the latitudes of 46° through 54°, listed on the right edge. The "Azimuth" scale, so labeled along the lower edge of the diagram, is divided every five degrees from 0° to 30°, then divided every degree from 30° to 126° and labeled 10° to 120° by tens. The hours, which are laid out along the top of the diagram, are divided into half hours from XII to I, and from I to VIII they are divided into 20-minute intervals and pricked

★ STAR LIST

Star name	*Declination*		*Mag.*	*Modern name*
Fomahanti	31	17	1	Alpha Piscis Australis
Os Pegasi	08	25	3	Epsilon Pegasi
Caput Delph	15	01	3	Alpha Delphini
Cor Vult	08	04	2	Alpha Aquila
Arcturus	20	56	1	Alpha Bootis
Caud Leonis	16	23	1	Beta Leonis
Cor Leonis	13	31	1	Alpha Leonis
Canis Minor	06	02	1	Alpha Canis Minoris
Med in cin Ori	01	19	2	Epsilon Orionis
Ocul Tauri	15	49	1	Alpha Tauri
Lucida Pleiad	23	04	4	Pleiades
Prim corn Ari	17	20	3	Beta Arietis

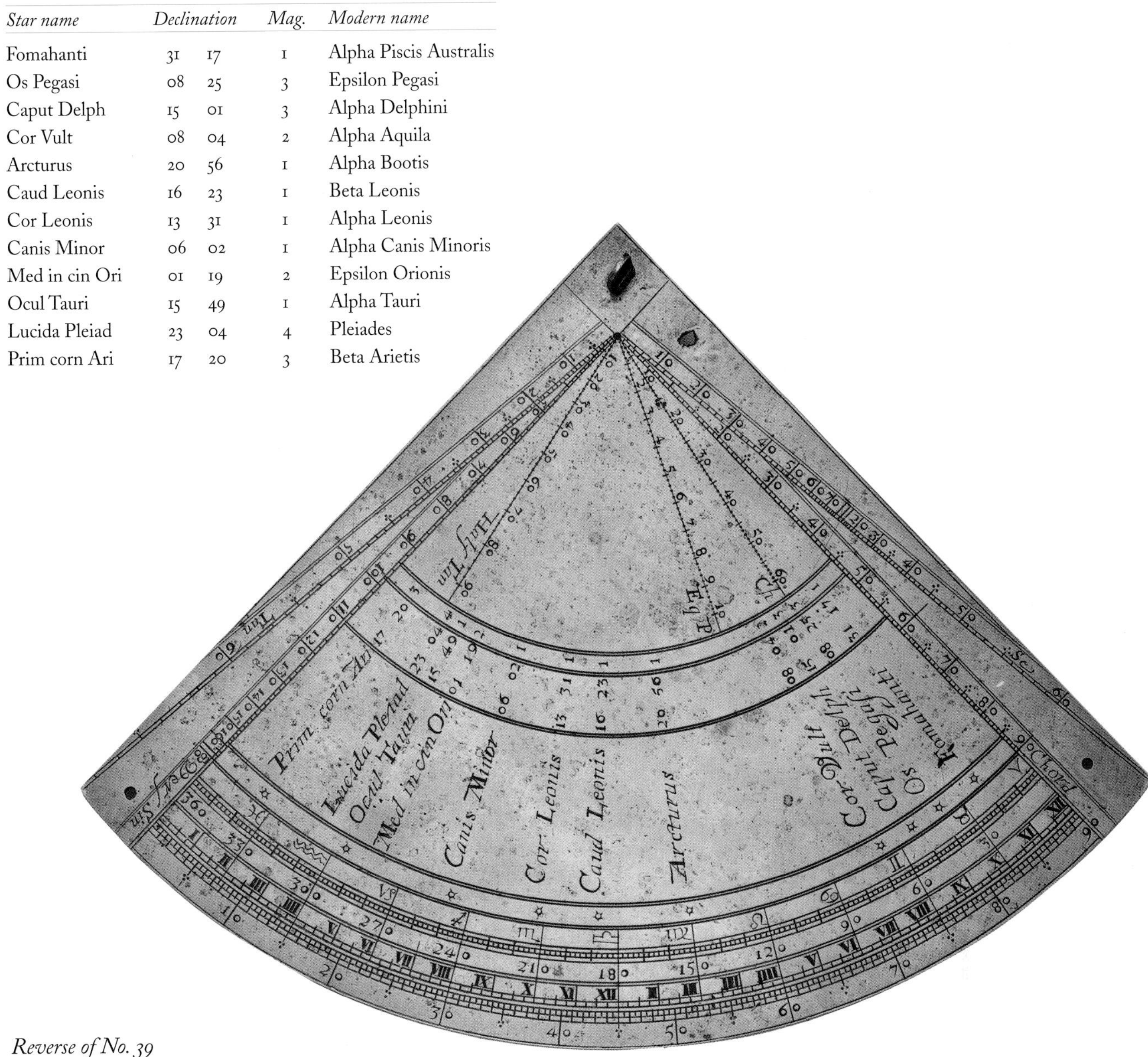

Reverse of No. 39

every fourth minute. The hour scale is labeled XII to VII and II to 5, both counter-clockwise. "Hour" appears on the upper left end. The latitude lines are pricked every degree and labeled 46°, 50°, and 54°.

The "Line of VI or 60" runs from the apex of the quadrant to the calendar. It intersects the analemma at the VI hour mark and at the 90° azimuth line. It crosses the zodiac at the point of Aries and Libra and ends at the civil calendar on March 10.

Above the diagram on the right side is inscribed "W. Hayes fecit," with "Dublin 1740" added later, surely by D. B. Sheahan. However, we can find no evidence in Mollan,[5] Burnett and Morrison-Low,[6] or Taylor[7] to back up this Irish attribution.

Inside the quadrant, along the left radius, is drawn a "line of three hours," which is a tangent line of 45°. The hour positions are marked by small stars and the line is divided, unequally, into quarter hours.

Outside the quadrant, in the wings, are four lines commonly found on a sector. On the left wing, the lines represented are one of cubes or solids and a second of numbers or equal parts. The latter is labeled "Lin." On the right wing, on the outside, is the line of squares or superficies, labeled "Sq," and on the inside the line of right sines, labeled "Sin." There is a battery screw near the apex on the left side. It appears to be extraneous.

On the reverse, the degree scale on the limb is divided every half degree, marked every five degrees, and labeled 10° to 90°. Inside this is the hour scale, divided every fifteen minutes and labeled I to XII twice. This represents the Right Ascension in time, with the first XII indicating midnight. The next scale inward is the Right Ascension in degrees, divided every two degrees, marked by tens, and labeled every 30 degrees, 360° to 0°, left to right. This scale is shared with the zodiacal calendar, labeled with the appropriate signs.

Next there is a radial list of stars arranged according to their Right Ascension, most preceded by a star symbol and followed by its declination and magnitude (★).

The declinations are somewhat different from Leybourn's, and we feel that Walter Hayes was working from a later corrected list or from a different one.

Above the magnitude arc, three lines radiate from the apex. The right-hand one is for chords, labeled "Ch" and running from 1 to 60. The middle one is for equal parts, labeled "Eq. P" and running from 1 to 10. The left-hand one is a tangent line, labeled "Half Tan" and running from 1 to 90.

The radii or wings carry four lines normally found on a sector. The outer one on the right is the line of natural sines and secants, labeled "Se." The inner one is labeled "Chord." On the left wing, the outer line is for natural tangents, labeled "Tan." The last line is for versed sines, labeled "Vers Sin."

We believe there were originally three sights, one at the apex and the others at the extremities of the wings. The one remaining sight is held on by an inappropriate nut. The plumb-bob and string are missing. This type of quadrant did not require a bead.[8]

We know of four other Panorganons, all unsigned. One is in the Physics Department at St. Andrew's University in Scotland. Another, of boxwood, is in the Bridewell Museum in Norwich.[9] The third, which is incomplete, is at the Whipple Museum of the History of Science at Cambridge.[10] The fourth known to us is at the National Maritime Museum at Greenwich.[11]

This instrument has many uses laid out and explained by Leybourn, and we will give only two useful instructions. In order to find the time of day, you must first measure the altitude of the sun using the quadrant. This is done by sighting the sun and noting where the thread crosses the scale on the limb. Next, measure the distance from the start of the sine scale to the number corresponding to the altitude found by using a pair of dividers. Then, having reset the thread to the day of the month, without changing the dividers mark off the distance from the thread along the proper latitude line. The other point of the dividers will mark the hour.

To find stellar time from the altitude of a star, the process is the same. You must use a star from the list on the reverse of the quadrant. To convert stellar time to solar time you must compare the Right Ascension of that star to that of the sun. These Right Ascensions are also found on the reverse of the quadrant.

This information plus descriptions of many other uses is to be found in the *Panorganon: A Universall Instrument* by William Leybourn. The edition we used was published in London in 1672. Prior to this, in 1658, John Collins wrote a treatise anticipating this instrument, which he called *The Sector on a Quadrant.*

I ARIES = 10 March.
PROVENANCE — John Tomlinson, Sr., 1927; John Tomlinson, Jr., 1927; A. P. purchase, 1936.
EXHIBITIONS — Metropolitan Museum of Art, New York, 1927-35; American Museum of Natural History, New York, 1935-36.
REFERENCE — Tomlinson (*c.* 1932).

1. Leybourn (1672).
2. Rees (1819), vol. 2.
3. Oxford Universal Dictionary (1955), 61.
4. Leybourn (1672), 1.
5. Mollan (1990).
6. Burnett and Morrison-Low (1989), 14.
7. Taylor (1966), 239.
8. Leybourn (1672), 2 in "To the Reader."
9. Holbrook *et al.* (1992), 188, 202, fig. 113.
10. Bryden Catalogue (1988), no. 290.
11. National Maritime Museum Catalogue (1970), 2: sect. 26-6.

40 Astrolabe-quadrant

Gunter-type, variant
John Prujean
Oxford
c. 1680
"Iohñ. Prujean Fecit Ōxō."
on the right end of the face
11.2 x 11.2 x 0.1 cm.
Working radius — 10.5 cm.
Brass
A-203

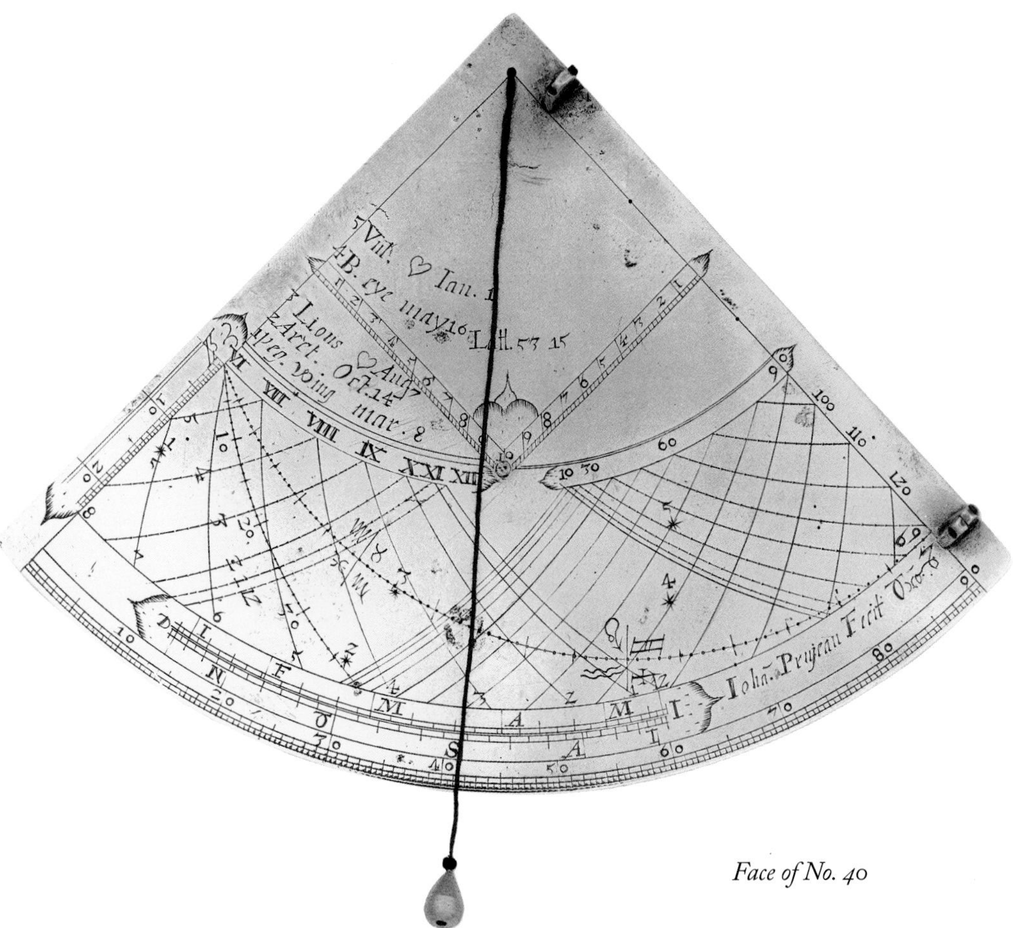

Face of No. 40

The face carries a typical Gunter-type quadrant[1] for latitude 53° 15'. The winter hours are shown by dotted lines and labeled VI to XII at the top and 5 to 12 (in reverse) below. The summer hours are labeled below 8 to 12 (in reverse). The azimuths are labeled 10°, 30°, 60°, 90°, 100°, 110°, and 120°. The declination scale is divided into half degrees and labeled by ten degrees. The shadow square is divided into 50 parts on each side and labeled 1-10-1. "Latt.53 15" is inscribed across the angle.

The dates indicate when the named stars will be in the south at midnight (★). They are also located by number between the equator and the Tropics.

On the calendar, the months are labeled with initials, with "J" being represented by "I." The months are divided into five-day intervals, with the last interval in each month representing one, three, or five days as appropriate. The arc scale is divided every half degree, marked by ones, fives, and tens, and labeled every ten degrees.

There are decorative lines with shading at the ends of the declination scale, azimuths, and calendar and in the corner of the shadow square. The line of the horizon is labeled 10°, 20°, 30°. The quadrant is signed to the right of the calendar: "Iohñ. Prujean Fecit Ōxō." The sights are present, and the plumb-bob and bead may be original.

The reverse has an "N" engraved below the hole for the plumb-bob line at the apex. A double circle is marked every quarter hour and half hour and labeled XII to XII twice. Within this hour circle rotates a volvelle with a rule. The outer scale on the volvelle carries the civil calendar, with the months labeled with initials; months beginning with "J" are represented by "I." The months are divided into five periods, with the appropriate number of days included in the last period of each

★ STAR LIST

5	Vul. ♡	Ian. 1
4	B. eye	May 16
3	Lions ♡	Aug. 7
2	Arct.	Oct. 14
1	Peg. Wing	Mar. 8

■ CONSTELLATIONS

[Ursa Major]
[Ursa Minor]
[Cassiopeia]
[Cepheus]
[Draco]

Reverse of No. 40

month. Inside the calendar there is a scale running clockwise from 23.5° at the left end of the Tropic of Capricorn, 23.5°-0°-90°-0°-23.5° to the right-hand end of the Tropic of Capricorn. It is divided every two degrees and labeled every ten degrees, starting with 20° on the left.

There is a typical de Rojas projection between the Tropics. Every tenth parallel or line of declination is accented by a dotted line. The ecliptic signs are drawn. The meridians are drawn every hour and are labeled 2 to 11 along the Tropic of Cancer and 11 to 1 along the Tropic of Capricorn, and below the "4" a later hand has added "IIII." Above and below the de Rojas projection, five circumpolar constellation figures are engraved, showing the principal stars in each (■). These figures are continued onto the volvelle using dotted lines and smaller stars. These are the constellations as depicted by Edmund Gunter in his drawing of the rundle, or volvelle, for a nocturnal.[2] There are five numbered pointers around the edge of the volvelle, corresponding to the list of stars on the face of the quadrant, with each at its appropriate date.

The rule is double and straight. The scale is divided every five degrees and labeled by tens, 90°-0°-90°, along the fiducial edge.

PROVENANCE — Found on an English beach; Phillip, Son, & Neale auction, London, 1971; A. P. purchase from London collector, 1971.
REFERENCE — Phillip, Son, & Neale Catalogue (1971), lot 135.

1. Gunter (1662), 104.
2. *Ibid.*, 64.

41 Astrolabe-quadrant

Gunter-type, variant
John Worgan
London
c. 1700
"I: Worgan fecit" *on the face*
20.4 x 20.4 x 0.15 cm.
Working radius — 19.8 cm.
Overall thickness, including gnomon — 9.0 cm.
Brass
W-88

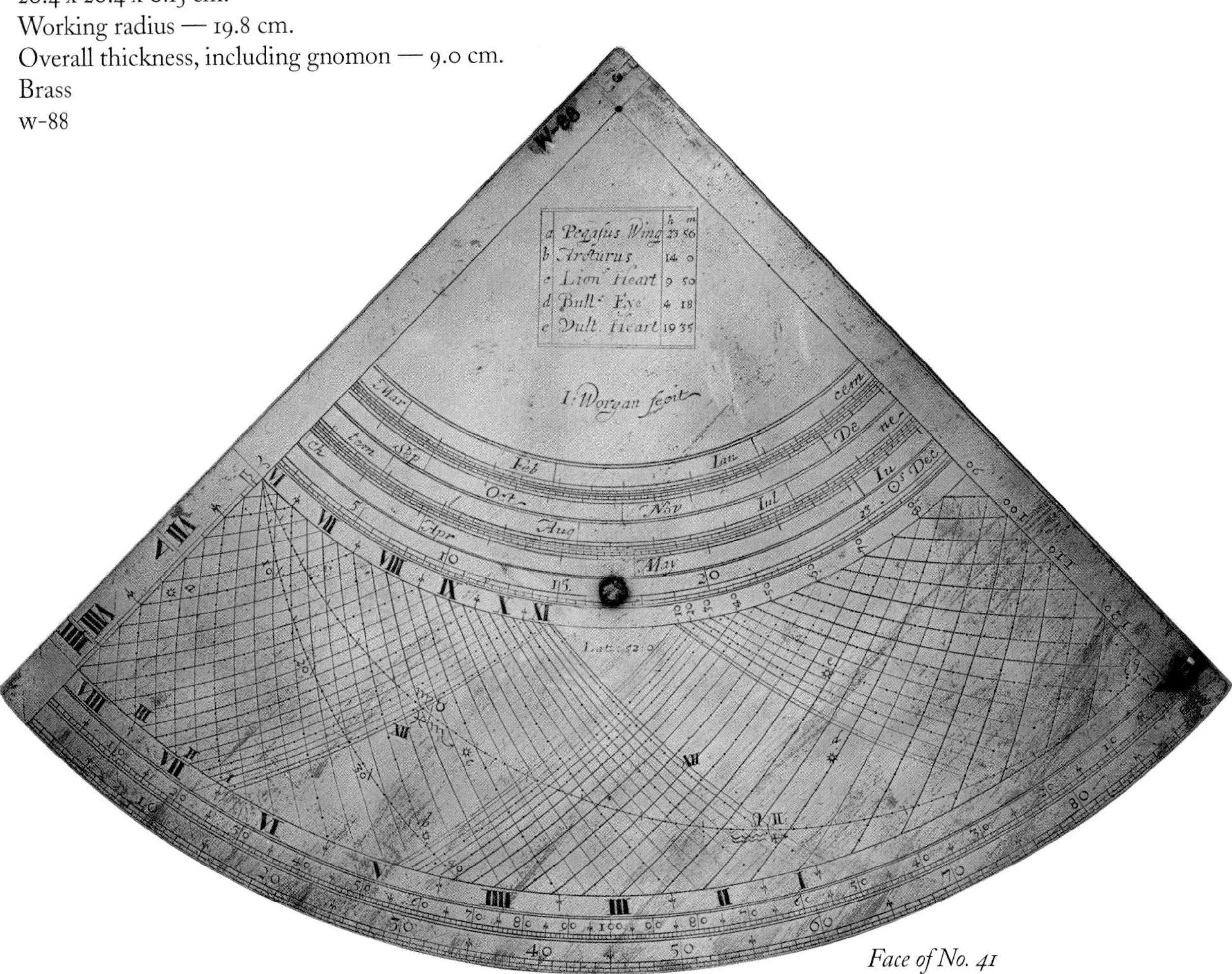

Face of No. 41

The face is laid out with a variant Gunter-type projection for the latitude of 52°. The limb carries the usual degree scale. The next inner scale, which is for the shadow square, is divided into parts, marked every five and ten parts, and labeled by tens, 10-100-10. The winter hours are labeled IIII and V along the left radius and VI to XI above the hour lines, with the XII line labeled between the horizon and the ecliptic. The summer hours are labeled VII and VIII along the left radius and then VIII to I (in reverse) along the Tropics line, with XII at the right-most summer line. The numeral four is shown as "IIII." The half hours appear as dotted lines. The azimuths are drawn every five degrees and labeled every ten, 10° to 120°. Every ten degrees, starting at 5°, is a dotted line. Between the hours and the azimuths, "Lat: 52:" is inscribed.

The declination scale is over the hour and azimuth lines and under the calendar. The arc is divided into half degrees, marked every five and ten degrees, and labeled 5°, 10°, 15°, 20°, and 23°. At the right-hand end, "☉s Dec." is engraved.

In use, while the string is stretched over the proper date and declination, the bead is set on the line of the ecliptic. The sun's altitude is measured, and the time is read from the position of the bead over an hour line.

Above the declination scale, the calendar is divided into four arcs. It starts at March 11 and runs alternately counter-clockwise and clockwise. Each arc represents a season — winter, autumn, summer, and spring — from the top down. The month names are divided at the ends of each arc.

The list of stars, with their Right Ascensions expressed in hours and minutes, is in a box below the apex (●).[1]

Below the box is the signature: "I: Worgan fecit." The position of each star on the quadrant is shown

RIGHT ASCENSIONS

	Star	Hr.	Min.
a	Pegasus Wing	23	56
b	Arcturus	14	0
c	Lion^s Heart	9	50
d	Bull^s Eye	4	18
e	Vult: Heart	19	35

Reverse of No. 41

by an asterisk and the letter representing that star in the box.

A replacement rivet of iron is placed on the midpoint of the declination scale. It holds the volvelle in place on the reverse. The upper sight and plumb-bob are missing.

The reverse has a horizontal sundial mounted on a volvelle to act as a declinatory. When the quadrant is aligned with a wall and the sundial is rotated until it reads the correct time, a pointer on the volvelle will indicate the declination of the wall on a circle of degrees engraved on the quadrant. This circle is divided every degree, marked every five degrees, and labeled every ten degrees, 0°-90°-0°, twice. The 45° mark is at the top of the quadrant.[2] Knowledge of the declination is a prerequisite for the design of an accurate wall dial.

The sundial on the volvelle is labeled "lat: 52:" — perhaps for Oxford. The hour scale reads VIII-XII-IIII counter-clockwise. Each hour has eight divisions, with the half-hour marks extended. There is a half sunburst engraved at the VI-VI line. A slotted plate is attached to the volvelle and carries a removable gnomon, which is cut out in a decorative curve and has a plumb-bob within the cut-out area. There is a leveling mark on the base of the gnomon.

At the apex a Tudor rose is engraved with flowing leaves. Each corner of the limb has a graceful tulip with its foliage. This is a typical decorative element used by Worgan.

I ARIES = 11 March.
PROVENANCE — New York State dealer, 1965; R. S. Webster, Winnetka, Ill., 1965; A. P. gift, 1983.

1. Gunter (1662), 128, first appendix to book 3, chap. 8.
2. Turner, A. J. (1987), 97.

42 Astrolabe-quadrant

Gunter-type
Daniel Harris
London
c. 1735
"D. Harris LONDON.*"*
on the face
16.0 x 15.3 x 0.7 cm.
Working radius — 14.4 cm.
Brass
DPW-41

★ STAR LIST

* PW	[Pegasus' Wing]
* Ar	[Arcturus]
* Lh	[Lion's Heart]
* Be	[Bull's Eye]
* Vh	[Vulture's Heart]

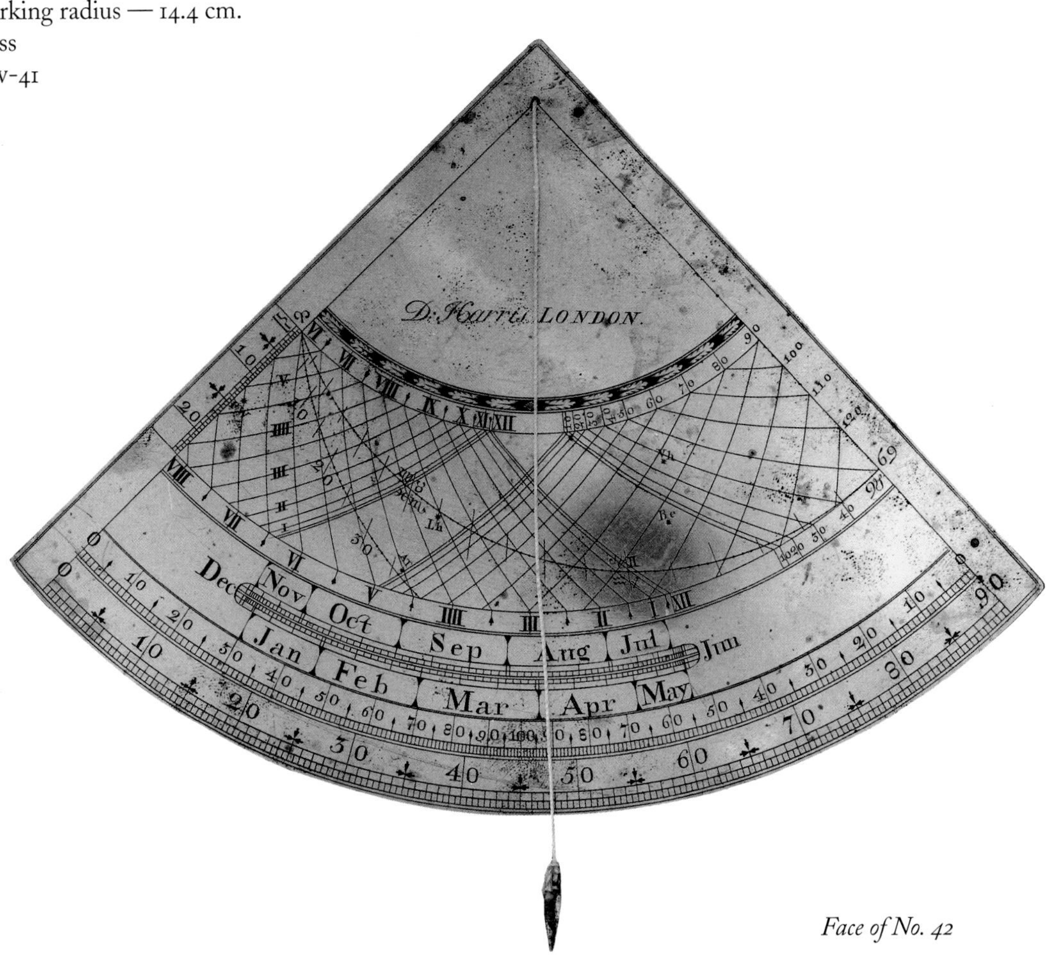

Face of No. 42

This Gunter-type quadrant is laid out for the latitude of London. The winter hours are labeled VI to XII at the top and V to I (in reverse) across the middle. The summer hours are labeled VIII to XII (in reverse). Both fours are shown as "IIII." The outer limb carries a degree scale divided into half degrees. Every ten degrees, starting with 5°, is marked with a cross, and it is labeled every ten degrees, 0° to 90°. The azimuths are drawn and labeled every ten degrees, 10° to 120° above and 10° to 40° below.

Within this arc lies the scale of the shadow square, divided by degrees, marked every ten degrees, starting at 5°, with an arrow, and labeled 0°-100°-0°.

The calendar is laid out above the shadow square, with the months labeled December to June and back again, June to December, showing the declination of the sun on any date, as is usual on Gunter-type quadrants.

The usual five stars are shown on the face of the quadrant with star symbols and initials (★).

Just above the hour and azimuth lines there is a decorative wheat design. The inscription lies below the apex: *"D. Harris* LONDON.*"* Daniel Harris was apprenticed to Edmund Blow of the Joiners' Company in 1723. He was turned over to Thomas Cooke I in 1725 and made free of the Company in 1735.[1]

The two brass sights are present, as is the plumb-bob, which is probably original. The reverse is blank.

PROVENANCE — David P. Wheatland, Topsfield, Mass., no. 1003; A. P. gift, 1986.
EXHIBITION — Hidalgo County Historical Museum, Edinburg, Tex., Aug. 17, 1992 - July 6, 1994.

1. Crawforth (1987).

43 Astrolabe-quadrant

Gunter-type
England
After 1752
Unsigned
"NEW STYLE"
on the right of the lower arc
12.8 x 13.2 x 1.7 cm.
Working radius — 12.4 cm.
Boxwood
DPW-46

★ STAR LIST

Star	*Date*	*Hr.*	*Min.*
Peg wi	Mar. 8	23	54
Arct	Oct. 14	13	58
Lio He	Aug. 7	9	48
Bul Ey	May 16	4	15
Vul He	Jan 1	19	33

● RIGHT ASCENSIONS

Star	*Hr.*	*Min.*
PW	23	56
Arc	14	00
LH	09	50
BE	14	18
VH	19	35

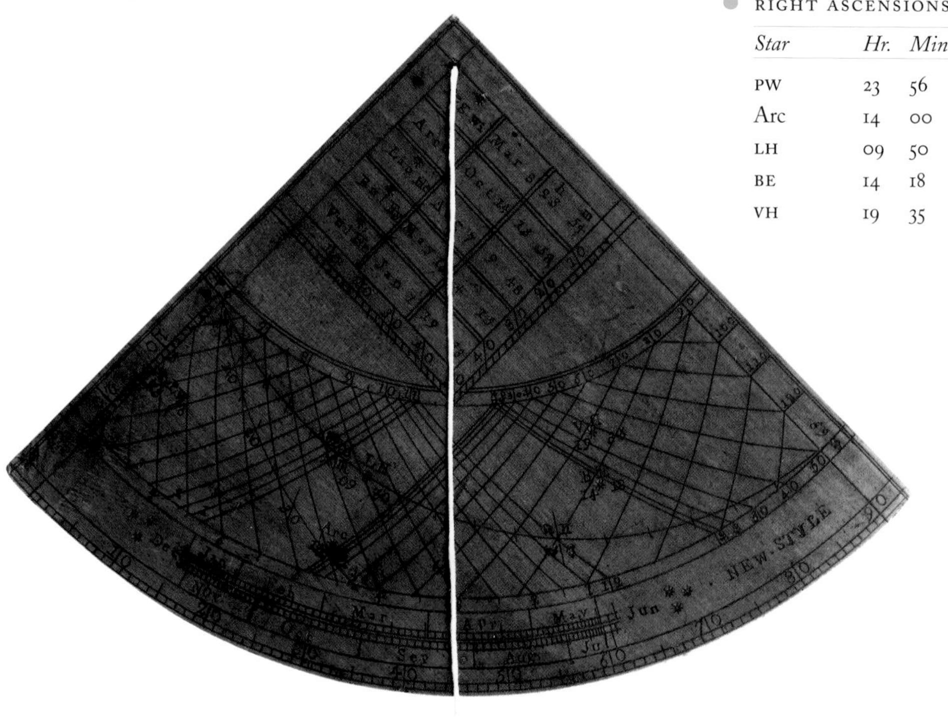

Face of No. 43

This quadrant has a Gunter-type projection on the face for the latitude of London. The winter hours are labeled 6 to 11 at the top and 3 to 1 (in reverse) below. The summer hours are labeled 8 to 12 (in reverse). The azimuths are drawn for and labeled every ten degrees, 10° to 90° at the top, 10° to 50° along the Tropics, and 100° to 120° along the right radius. The declination scale is divided and labeled as usual. The line of the ecliptic is labeled 0°, 20°, 30°, and 40°. The shadow square is divided into 50 parts on each side and labeled by tens, 10-50-10. Within the shadow square, a lined chart is laid out along the left radius. It shows the usual five stars, with the date of their south positions at midnight and their Right Ascensions given in hours and minutes (★).[1]

These same five stars, with their times, are shown on the quadrant between the equator and the line of the Tropics (●). The first three stars and the last show two minutes more than the list in the shadow square, and the fourth shows ten hours and three minutes more.

The sights are missing, and the plumb-bob and line are replacements. The right radius has a hole to take the plumb-bob. The bead is missing. The reverse is blank.

The inscription, "NEW STYLE," indicates that it was made after 1752, when England adopted the Gregorian calendar.

PROVENANCE — David P. Wheatland, Topsfield, Mass., 1926, no. 186 = no. S. 363; A. P. gift, 1986.

1. Gunter (1662), 128.

44 Astrolabe-quadrant

Gunter-type
London
After 1752
Unsigned
10.4 x 10.1 x 0.8 cm.
Working radius — 9.4 cm.
Boxwood
W-70

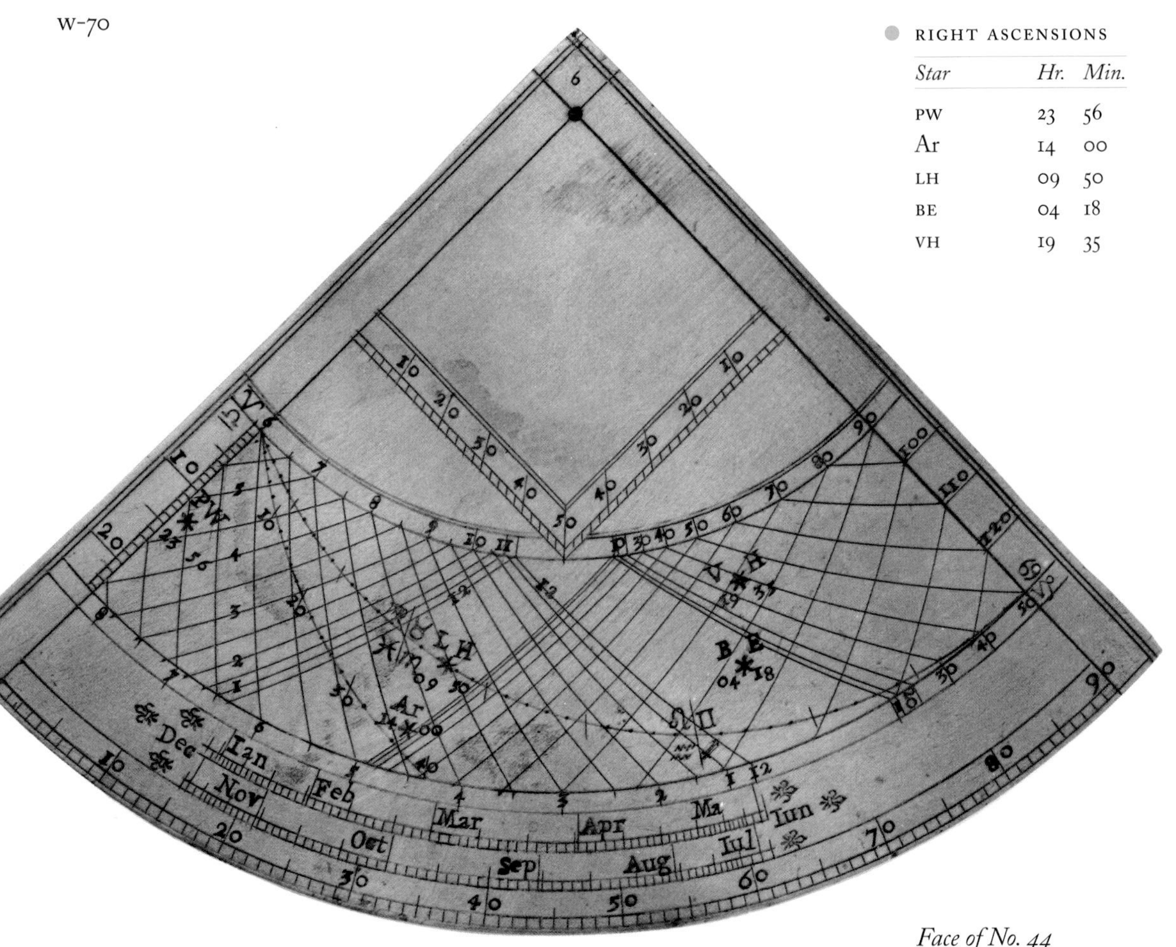

● RIGHT ASCENSIONS

Star	Hr.	Min.
PW	23	56
Ar	14	00
LH	09	50
BE	04	18
VH	19	35

Face of No. 44

This quadrant has the usual Gunter-type projection for the latitude of London. The winter hours are labeled 6 to 11 at the top, and the summer hours are labeled 8 to 12 (in reverse) along the bottom. The azimuths are labeled by tens, 10° to 90° along the top edge, 100° to 120° along the right radius, and 10°, 30°, 40°, 50° along the bottom. The limb is divided every degree and labeled as usual. The calendar is also labeled as usual, with "J" being written as "I." The declination scale is shown as usual.

The shadow square is divided into 50 parts on each side and labeled 10-50-10. The usual five stars are laid out on the face, each with a star symbol and its Right Ascension in hours and minutes (●). There is a hole for the plumb-bob at the apex with a "6" stamped above it.

There is also a hole along the left radius for the plumb-bob, which is missing. The sights are also missing.

The reverse has a sight and a plumb-bob drawn in pencil. There is also a list of stars, their Right Ascensions, and the dates on which they cross the south meridian at midnight (★). This table is also written in pencil.

PROVENANCE — R. S. Webster, Winnetka, Ill., 1970; A. P. gift, 1983.

★ STAR LIST ON REVERSE

Star	Date	Hr.	Min.
Peg Wi	March 8	25	14
Arctu	Oct. 14	15	18
Lio Ha	Aug. 7	9	48
Bul Ey	May 6	4	11
Vul Ha	Jan 1	19	33

45 Astrolabe-quadrant

Gunter-type
London
After 1752
Unsigned
15.3 x 15.0 x 1.0 cm.
Working radius — 13.8 cm.
Boxwood
DPW-12

● RIGHT ASCENSIONS

Star	*Hr.*	*Min.*
PW	23	56
Arc	14	00
Lh	09	50
Be	04	18
Vh	19	35

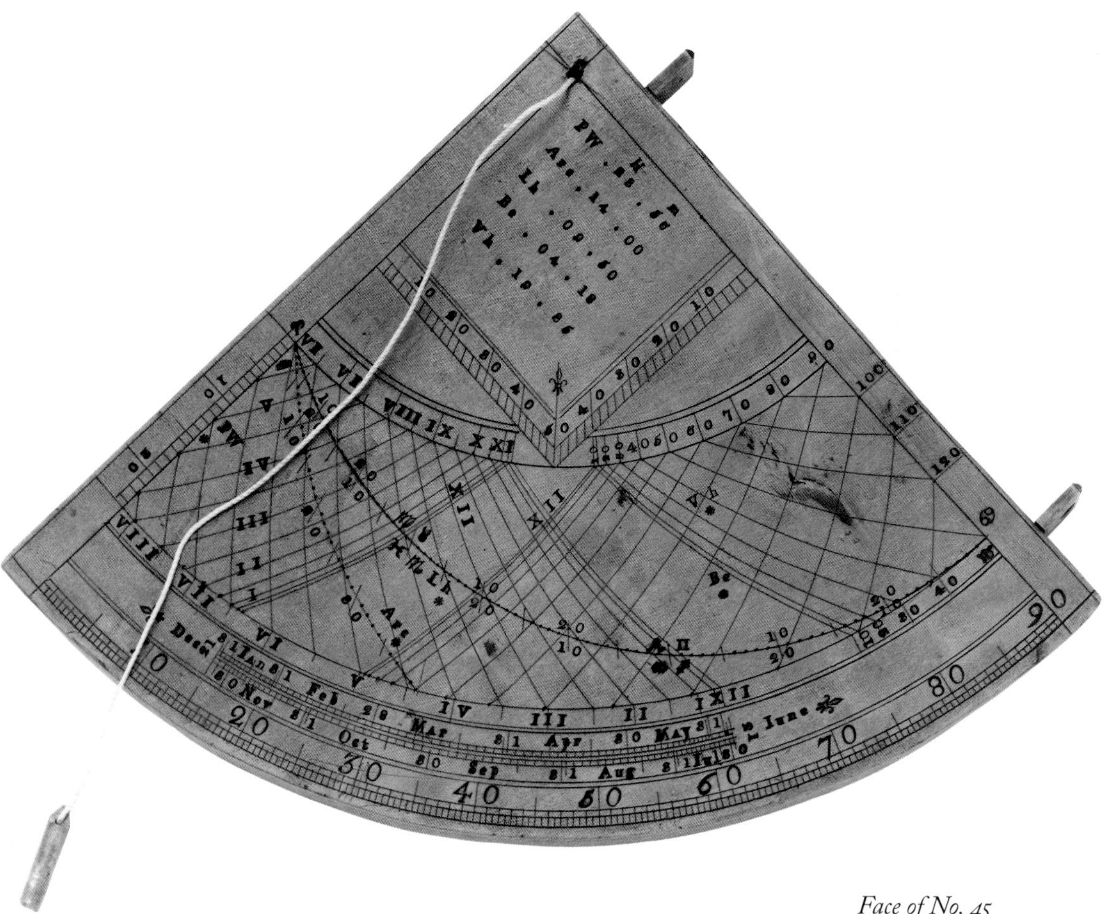

Face of No. 45

The face of this quadrant carries a standard Gunter-type projection for the latitude of London. The winter hours are labeled VI to XII above and V to I (in reverse) below. The summer hours are labeled VIII to XII counter-clockwise, above the Tropics. The azimuths are labeled every ten degrees, 10° to 120° at the top and right radius and 10° to 50° along the bottom. The line of the ecliptic is divided every degree, marked every ten degrees, and labeled to show the zodiacal signs. The horizon is also marked every degree and labeled 10°, 20°, 30°. Both lines have longer marks every five degrees. The declination scale is divided and labeled as usual. The calendar is labeled as usual, with "Dem~" being the only variant spelling. The 45° line falls on April 13.

The shadow square is divided every two parts, marked every ten parts, and labeled 10-50-10. A fleur-de-lys is in the corner. Within the shadow square, the usual five stars are listed along the left radius, with their Right Ascensions given in hours and minutes (●).

The same stars are shown on the face, marked with star symbols.

The plumb-bob is original. It fits into a hole on the left edge for storage. The brass sights are on the right edge. The reverse is blank.

PROVENANCE — David P. Wheatland, Topsfield, Mass., no. 368 = no. 365; A. P. gift, 1986.

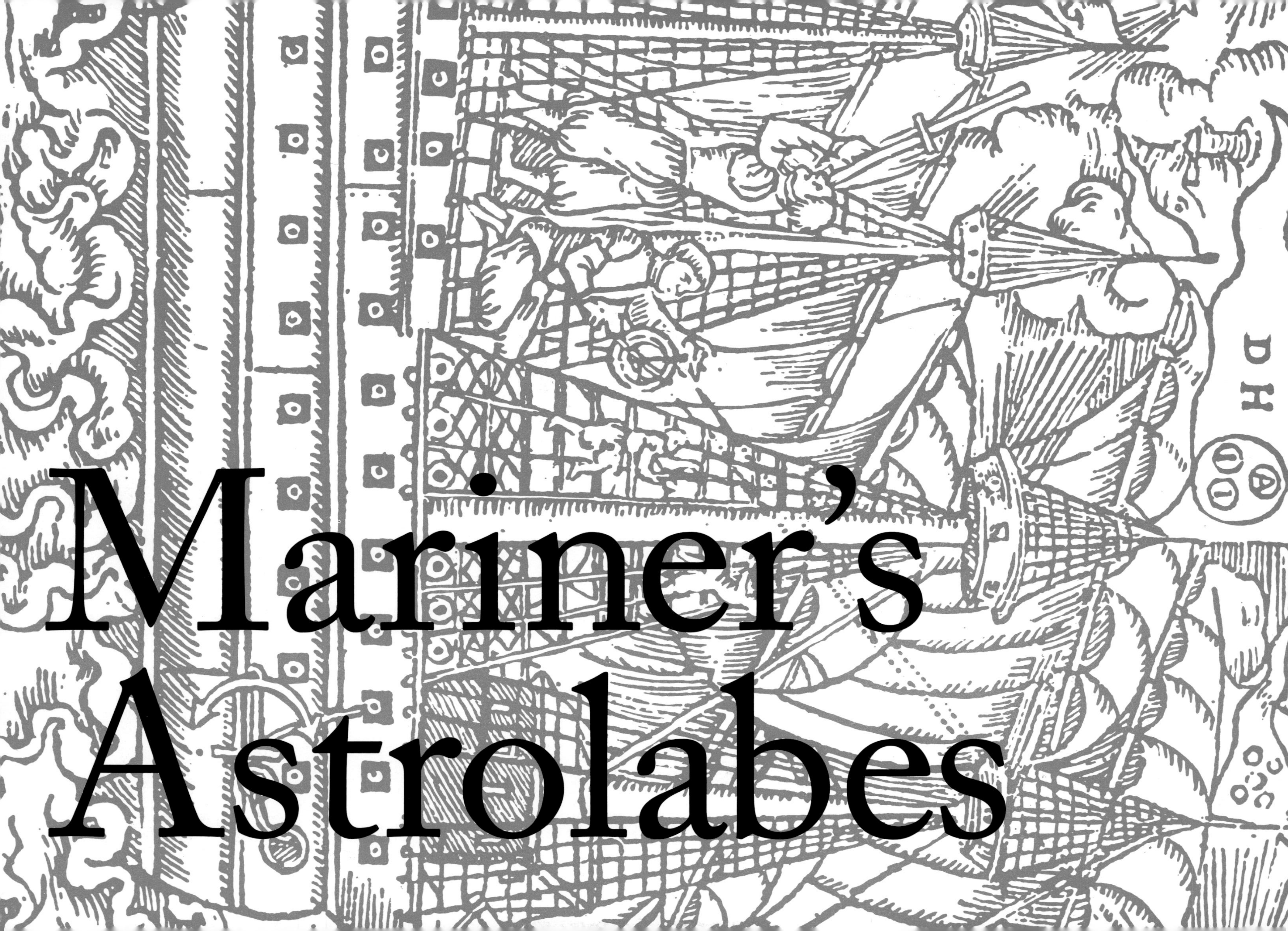
Mariner's
Astrolabes

46 Mariner's astrolabe

Portugal
1616
Unsigned
21.9 x 17.1 x 8.9 cm.
Thickness at top — 2.1 cm.; thickness at bottom — 2.2 cm.
Weight — 3,013 g.
Brass
A-275

Face of No. 46

This mariner's astrolabe is a cast wheel type with base ballast, Stimson's type 1(a).[1] It is similar to other known Portuguese astrolabes in its design, punchmarks, and scale. The surface is pitted. The throne was cast with the body and consists of two low scrolls. The bail and ring are pivoted.

The limb has a scale divided every degree, marked and labeled by fives on the outside and by tens on the inside, 90°-1°-90°. It thus measures the zenith distance of the sun or a star. Scales showing 1° instead of 0° at the top or bottom are found on others dating from 1602 to 1623.[2] The reverse is blank.

The date "1616" is punched on the rounded support at the bottom (the ballast), with four punchmarks around it. These are diamond-shaped, four-lobed rosettes with a circle in the center. There are seven punchmarks altogether: four around the date, one inside each 90° mark on the limb, and one at the top, under the 1° label. There is also a "1" punched at the bottom where the 0° label would have been if the scale had been continued around the astrolabe.

The alidade is counterchanged, with the pinhole sights close together. The ends come to long, narrow points. The original bolt, which has a knob head and was hand-threaded, and the original wing nut are present.

This mariner's astrolabe was one of five found on the Spanish ship *Nuestra Señora de Atocha.* This was one of four found in or near the pilot's chest. The *Atocha* was bound for Spain when it ran into a hurricane and sank near the Marquesas Keys, off the coast of Florida, in September 1622. Attempts by the Spanish to salvage her failed when another hurricane in October broke up the ship.

This astrolabe is no. 58 in the National Maritime Museum Registry of Mariner's Astrolabes[3] and Atocha II (lot 20) in Christie's catalogue.[4]

PROVENANCE — *Nuestra Señora de Atocha,* Spain, 1616-22; Treasure Salvors, Inc., Key West, Fla., 1986; Trevor Philip and Son, London, 1988; A. P. purchase, 1988.

REFERENCES — Adler (1989), cover and inside cover page; Christie's New York (1988), 62-63, lot 20; Stimson (1988), 164-66; Turner, G. L'E. (1991), 234 (plate); Waterman (1997), 38-39.

1. Stimson (1988), 164.
2. *Ibid.*, 56.
3. *Ibid.*, 164-66.
4. Christie's New York (1988), 62-63, lot 20.

47 Mariner's astrolabe

Laurits Christian Eichner
Bloomfield, New Jersey
1963
"1602"
on the face
"1963 / L.C.Eichner"
on the back
22.4 x 17.0 x 7.2 cm.
Thickness at top — 1.5 cm.; thickness at bottom — 1.9 cm.
Weight — 2,381.4 g.
Brass
A-157

This is a copy of the Barlow astrolabe,[1] which is at the National Museum of American History. The throne has the shape of a double scroll; both the bail and the ring are pivoted. The alidade is equipped with pinhole sights mounted 6.6 cm. apart. The limb is divided every degree and marked and labeled every five degrees, 90°-1°-90°, above the right horizon.

LEFT: *Face of No. 47*
RIGHT: *Reverse of No. 47*

PROVENANCE — A. P. purchase, 1965.
REFERENCES — Multhauf (1971), 39; Stimson (1988), 70-71.

1. Waters (1966), 32; Stimson (1988), 70-71.

Astrolabium

Appendices

Star Catalogue

For the exact form of each star name, please refer to its catalogue entry.

ua = universal astrolabe
sq = shadow square
g = grid
r = reverse

ANDROMEDA		Cat. No.
Alpha	Alfraz	1
	Cap Andromede	4
	Vmb, Andro,	5
	Pegasi vmbi:	8
	Pegasi vmbil:	9
	caput anDromaDa	11
	Alpharez	18
	Caput Andr	25
	Vmbilic Andron	28
Beta	Mirac	1
	Medi: cing: andro:	8
	Andro: vmbilic[9]	9
	anDro	11
	Andromeda	19, 31
	Cing. Andromedae	24
	Cin Andr	38
Gamma	Andoromade: achr	16
?	Andro Scapu:	16
ANTINOÜS		
Alpha	Prima Antinoi	18, 24
AQUARIUS		
Delta	Schcack	1
	Crus aquarij	8
	Crus ≈	9
	Crus aquarium	10
	crus aquarius	11
	Avst	14
	Scheat	18
	Crus Aquary	24
	Crvs	28
Epsilon	Clara D Humeri ≈	18
AQUILA		
Alpha	Altair	1, 30, 32, 35
	althair	2, 3
	Aquila Volans	4, 24
	Aquila	5, 8, 8*ua*, 9, 9*ua*, 10, 16*g*, 19, 27, 29, 31
	aquilo	10*ua*
	vvltuvo	11
	Aqvila V·	14
	Aq.	17
	Aguila	26

AQUILA *(cont.)*

		Cat. No.
	Vvltvr. Vol.	28
	Altar	33, 34
	VH	37, 42, 43, 44, 45
	Cor Vult	38, 39
	Vul. ♡	40*sq*
	Vult: Heart	41
	Vul He	43*sq*
	Vul Ha	44*r*
Epsilon	Vvlt Ca	28

ARGO NAVIS

Alpha Carinae		
	Canopus	8*ua*, 9*ua*, 10*ua*, 30
	Argo Naui	16
	Cano.	17
	Canop^{9}	26
	Canop	27
Chi Velorum		
	Markeb	1
?	Navis	4

ARIES

Alpha	Enf	1
	Cornu Arietis	16
	Cap Arie	38
Beta	cornu	2, 3
	Prim corn Ari	39

AURIGA

Alpha	Alhaioc	1
	alhatot	2, 3
	Hircus	8, 8*ua*, 9, 9*ua*, 10, 10*ua*, 14, 16, 19, 26, 27, 36
	Alaior	11
	Hir.	17
	Alhejoth	18
	Hirc^{9} Capel	24
	Capella	25, 30, 35
	Hircvs Ca	29
	Hircus s	38
Eta	Hadorum prece	16
?	Avriga	4

BOOTES

		Cat. No.
Alpha	Alramec	1
	alcameth	2, 3
	Arcturus	4, 5, 8, 8*ua*, 9, 9*ua*, 10, 10*ua*, 19, 24, 30, 31, 33, 34, 35, 38, 39, 41
	Artur boet	11
	Art	14, 29
	Arctu Bootes	16*g*
	Arct.	17, 40, 43*sq*
	Alramech	18
	Artvrvs	21
	Arctur^{9}	26
	Arctur·	27
	Bootes	28
	a	37
	Ar	42, 44
	Arc	43, 45
	Arctu	44*r*
Gamma	Bootis sinister hu:	8
	Bootis sinist: hum:	9
	Bootis sinister Humerus	10

CANCER

Gamma	Asyllus Austrinu	16
Delta	Asellus Boreus	16
?	cancer	11

CANIS MAJOR

Alpha	Alhabor	1, 2, 3, 18
	In ore canicvle sidvs magne Lvcis	4
	Canis	5
	Canis Maior	8, 8*ua*, 9, 9*ua*, 10, 10*ua*, 16, 19, 21, 24, 31
	Canis major	25
	canis ma	11
	Sirius	14, 27, 30, 32, 33, 34, 35
	C. ma.	17
	Can. mai.	26
	Syrivs	28
	Sirivs Cor	29
	Canis Maj	38

CANIS MINOR

		Cat. No.
Alpha	Algomerza	1
	algimeica	2, 3
	Procion	5
	Canicula	8, 8*ua*, 9, 9*ua*, 10, 28, 38
	Canis Minor	10*ua*, 16, 19, 21, 24, 25, 29, 31, 39
	Canis mi	11, 26
	Canis Minor Procioh	14
	C. m.	17
	Algomeisa	18
	can mi	27
	Procyon	30, 33, 34, 35
Beta	In Collo Caīs	4
?	Capvt Canis	4 (maybe Canis Major)

CAPRICORN

Alpha	Capra	5, 31
	Cornu Cap	38
Gamma	Cauda capricorni	8, 10
	Cauda ♑	9
	C· Capr·	14
	Cauda Capricor:	16*g*
	Capricornvs Av	29
Delta	Dheneb Algedi	1
	Cau. Cap.	17
	Cavda Capricor	19
	Cauda ♑	26

CARINA see Argo Navis

CASSIOPEIA

Alpha	Sceder	1
	Pect' Casiepie	4
	Pect^{9} cassi	8, 9
	Pectus cascio	10
	P· Cas·	14
	Pect^{9} Cassiop.	24
	Pect. Cassiop	28
	Pv Cas	29

CEPHUS

Alpha	Dherat	1
	Cephei dex: hum:	9

CEPHUS *(cont.)*

		Cat. No.
	cefeu umi	11
	Dex hu Ceph	16*g*

CETUS

Alpha	Menkhar	1
	memkar	2, 3
	Naris Ceti	4, 31
	Nares Ceti	5, 8, 9
	Ceti nares	10
	Ceti Juba	16
	Menkar	18, 33, 34, 35
	Os Ceti	38
Beta	Dheneb Caitoz	1
	Cauda Ceti	4, 5, 8, 9, 10, 16, 19, 31, 36
	Deneb Kaytos	18
	Cauda Ceti Aust:	24
	Cavd Ceti	28
	Diphda	33, 34, 35
Gamma	Os. Ceti	28
Zeta	Betin Caitoz	1
	Venter Ceti	2, 3, 4, 5, 8, 9, 10, 11, 14, 19, 21, 29, 31
	Ceti Venter	16
	Baten Kaytos	18
	Vent. Ceti	28

CORONA BOREALIS

Alpha	Elfeca	1
	elfeta	2, 3
	Corona	4, 5, 28, 31
	Corona sept:	8
	Corona Septent:	9
	Corna sept	10
	coro	11
	L· Co·	14
	Lucida Corp: Gnos	16*g*
	Alpheta	18
	Lco	29
	Alpheca	33, 34
	Corona Se	38

CORVUS

Alpha	Corvvs	5, 11, 31
	R· Corvi	14

CORVUS	*(cont.)*	Cat. No.
	Corui Rostr̃	16*g*
	Rostrvm Corvi	19, 36
Gamma	Algorab	1
	Corui ala dextra	8, 9, 16
	Coruiola dextra	10
	Ala· D· C·	14
	Algorab	18
	Ala D. Corv.	28
	Scorvi Ala	29
?	Decorvo	1
	Corvvs	4, 11

CRATER		
Alpha	Crater	5
	Crateris fund[9]	8
	Crateris fundus	9
	Cariens fundus	10
	Alkes	18
	Patera	31
?	In Basi Crateris	24

CRUX		
Alpha	Crvs Meridionale	4
	α Crucis	30

CYGNUS		
Alpha	Addigege	1
	ariof	2, 3
	Cavda Galine	4
	Holor	5, 31
	Cauda cijgni	8
	Cauda cÿgni	9
	cauDaga	11
	C· Cig·	14
	Cauda cigni	16*g*
	Cÿg.	17
	Cavda Signi	19, 21
	Caud cigni	24
	Cavda Gallinae	28
	Cig	29
	Deneb	30, 33, 34, 35
	Cauda cygni	36
	Cau Cygn	38
Beta	galina	11
	Rostr galinae	24

CYGNUS	*(cont.)*	Cat. No.
Gamma	Pectus Cygni	18
Epsilon	Hvm. Si.	28

DELPHINUS		
Alpha	dalPhi	11
	Delphin	5, 31 (5 could be Epsilon)
	Caput Delph	39
Epsilon	Delfin	1
	Cavda Delphin	19
?	Cauda Dephu	16*g*

DRACO		
Alpha	Caput Draco	36 (possibly Lambda)
	Dr	29
Beta	Cap, Serp,	5
	Cap Drac	38
Gamma	Taben	1
	Capvt Serpentarii	4
	Caput Draconis	8, 9, 10
	draco	11
	C· Dr·	14
	Caput dra	16*g*
	Ras Aben	18
	Cap: Serp:	31, 36
[Lambda]	Caput Draco	36 (more likely Alpha)
?	Pal:Si: Serp:	31

EQUULEUS		
Alpha	Eqvvs Prior	4

ERIDANUS		
Alpha	Postr: aquae fusae	9*ua*
	Postremasu: aque	10*ua*
	Extre: Eridani Acarn	16
	Stella Eridani	18
	Eridan	27
	Achernar	30
Theta	Post interualum fluuy	24

ERIDANUS *(cont.)*

		Cat. No.
Tau	Avgetavar	1
?	Eridanvs	4

GEMINI

		Cat. No.
Alpha	Raz	1
	Caput ♊ anter:	9, 9*ua*
	Apollo	16
	Cap. ♊ ant.	26
	Castor	30
Beta	Hercules	16
	C. ♊ ant.	17
	Pollux	30
?	Capvt Promiget Nor	4

HERCULES

		Cat. No.
Alpha	Caput herculis	8, 9
	Caput Hercules	10
	C· Her·	14
	Caput Engouna	16*g*
	Ras Algethi	18
	Cap Herculis	24
?	herculis ala	11
	Hir	29

HYDRA

		Cat. No.
Alpha	Alfard	1, 33, 34
	ydra	2, 3
	Hidra	5, 26, 31
	Hijdre clara	8
	Hÿdrae clara	9
	lucida Hÿdre	10
	idra	11
	Lvcida Idrae	14, 29
	Cor Hydre	16
	Luc. hÿ.	17
	Alphard	18, 35
	Lvcida Hidrae	19
	Lucida Hydrae	24
	Cor hidrae	25
	Hydra	28
	Cor Hydr	38

LEO

		Cat. No.
Alpha	Cor	1
	cor le	2, 3

LEO *(cont.)*

		Cat. No.
	Cor Leonis	4, 8, 8*ua*, 10, 14, 16, 19, 21, 31, 39
	Regulus	5, 30, 35
	Cor ♌	9, 9*ua*, 10*ua*, 24 25, 27
	Cor L.	17
	Kalb Eleced	18
	♡ ♌	26
	Rex [in a ♡]	28
	Leonis	29
	LH	37, 42, 43, 44, 45
	Cor Leoni	38
	Lions ♡	40
	Lion^{S} heart	41
	Lio He	43*sq*
	Lio Ha	44*r*
Beta	Cavda	1
	Cavda Leonis	4, 31
	Cavda Leon,	5
	Cauda ♌	8, 9, 9*ua*, 16, 19, 24, 25, 26
	Cau. Le.	17
	Cau ♌	27
	Cavd. Leon	28
	Denebola	30
	Caud Leonis	39
Gamma	Dor. L.	28
Delta	Dorsū leonis	8
	Dorsum ♌	9, 24
Theta	Ceruix leo	16
?	Pirm ♌	29

LEPUS

		Cat. No.
Alpha	Medium Leporis	18

LIBRA

		Cat. No.
Alpha	Lanx	28
	Libr Sc	29
Beta	Lanx boree cla:	8
	Lanx boree clarior	9
	Lancus Borea	16*g*

LYRA

		Cat. No.
Alpha	Wega	1

LYRA *(cont.)*		Cat. No.
	wegua	2, 3
	Lira	5, 19, 25, 27
	Lijra	8, 8*ua*
	Lÿra	9
	Lyra	9*ua*, 26, 36
	liera	10, 10*ua*
	v. ca	11
	Sp· Lirae Fidicvla	14
	Fidicula	16*g*, 29
	Lÿr.	17
	Alchar Fidicula	18
	Vega	30, 31, 33, 34, 35
	Lu Lyrae	38
Beta	Seli	29
?	Aqvila Cadens	4

OPHIUCHUS		
Alpha	Alhawe	1
	alhave	2, 3
	Ophiuc: cap:	8
	Ophiuchi caput	9
	Caput Ophiuchi	9*ua*, 16*g*, 19
	Ophiuci caput	10
	Ras Alhague	18
	Ophivch	28
	Rosalhague	33, 34
Beta	Ophiuchi ma: dex:	8
	Ophiuc: ma: dex:	9
Delta	Yed	1
	Man^9 Serp,	5
	Ophiuchi sinister ma:	8
	Ophiuc: sin: manus	9
	Palma Ophiuci	16*g*
	Palma Ophivchi	19
Zeta	Sinist genu	24
Eta	Ophiuchi genu dex:	9
	Ophiuci Dex. genu	24
?	serPentarius	11

ORION		
Alpha	Elgevze	1
	algerice	2

ORION *(cont.)*		Cat. No.
	Hvs Orionis	4
	Orion: dexter hũer^9	8
	Dex: hũe: orioĩs	8*ua*
	Orionis dex: hum:	9
	dester hume	11
	D· Hvm· Or	14
	hu: dex Ori	16
	D. h. Or.	17
	Bed Algeuze	18
	Dex Hvmervs Orio	19
	Orionis hum	25
	Dexte. Hu. orionis	26
	Hu. dex Or	27
	Hvm. D. Orion	28
	Proc. Med. Or. I	29
	Betelguese	30, 35
	Betelgeuse	33, 34
Beta	Rigel	1, 2, 3, 30, 33, 34, 35
	Pes Orionis	4
	Pes sin, Orionis	5
	Orionis sinis: pes	8
	Orio: sin: pes	9
	Orionis sinist pes	9*ua*
	Orionis sinister pes	10
	Orionis si: pes	10*ua*
	orionis	11
	Sin· Pes· Orionis Rigel	14
	Sinister ps: Orionis	16
	Sinister Pes Orion	19
	Pes Sinister Orionis	21
	Pes Orion	25, 38
	pes si. Or.	27
	Sin. Pes. Orionis Rigela	29
	Se'pes Orionis	31
Gamma	Orion: sinister hum:	9
	Orionis sinister Humerus	10
Delta	Orio	16
Epsilon	Cingu: orioni media	9
	Med· C· Or	14
	Cin	16
	media Cing Orio	24
	Med in cin Ori	39

ORION *(cont.)*		Cat. No.
Zeta	Cingu Orionis	16
?	Orion	28

PEGASUS

Alpha	Hvme, Eqvi	5
	Pegasi ala	8
	Ala pegasi	9
	Pegasi umbilicus	10
	Marchab Pegasi	18
	Pegasvs	31
	Mark	33, 34
	Markab	35
	pn	37
	PW	42, 43, 44, 45
	Peg. Wing	40
	Pegasus Wing	41
	Peg Wi	43*sq*, 44*r*
Beta	Hv Eqvi	1
	humer^{9}	2, 3
	Crus Pegasi	5, 9, 16*g*, 19
	Pegasi crus	8, 10
	Dex. hum. Pega	24
	Hv. Dex. Eq.	28
	Hvme: Eqvi	31
	Alfer	33, 34
Gamma	Eqvi Ala	28
	Ala Pega	38
Epsilon	Mv Eqi	1
	Mvsida Eqvi Pegasi	4
	Pegasi rictus	8, 9
	Pectus Pegasi	24
	Os Pegasi	39
Eta	Hvmervs Pegasi	4
	Pegasi Humerus	8, 9, 10

PERSEUS

Alpha	Algenb	1
	Lat^{9} Dextrv̄ Psei	4
	Pers: lat^{9} dex:	9
	persei	11
	Dex Lat Per	19
	Dex. lat^{9} Persei	24
	Persei Dex.	28

PERSEUS *(cont.)*		Cat. No.
Beta	Gorgon	5, 31
	Meduse caput	8, 9
	Caput meduse	8*ua*
	meDuse	11
	Ca Al	14
	Caput Medulse	16
	Algol	18
	Cap Med	19
	Cap. Algol	24
	C. Gorg.	28

PISCES

Alpha	Flvxvs	4

PISCIS AUSTRALIS

Alpha	Fomahand	16*g*
	Aquarii Fomahant	25
	Postr. fu. aque	26
	Aqua	27
	Fomalhaut	30
	Fomaha	38
	Fomahanti	39

PLEIADES

	Pleiadum (7 stars)	16
	Lucida Pleiad	39

SAGITTARIUS

Epsilon or Lambda		
	Cuspis Sagitari	16*g*

SCORPIO

Alpha	Corvs^{9}	1
	cor	2, 3
	Cor Scorpii	4, 19
	Scorpij cor	8, 8*ua*
	Cor ♏	9, 10, 14, 25, 27
	scorPio	11
	Cor Scorpi	16*g*, 24
	Cor sc.	17
	♡ ♏	26
	[image of scorpion]	28
	Cob ♏	29
	Antares	30, 35
	Cor Scor	38
Gamma	Aust	16*g*
	Lucida lancis Aust.	24

SCORPIO *(cont.)*		Cat. No.
Delta	Frons	16*g*
Lambda	Cauda ♏	9*ua*
Xi	Borea	16*g*
TAURUS		
Alpha	Aldebarn	1
	Aldebaran	2, 3, 18, 30
	Oculus Tauri	4, 5, 8, 16, 19, 21
	Oculus ♉	8*ua*, 9, 9*ua*, 10, 10*ua*, 25, 31
	taurus oculi	11
	Aldeb· Oc· Tavri	14
	Oc. t.	17
	Ocul^{9} Tauri	24
	Ocul^{9} ♉	26
	Ocul ♉	27
	H. Ocvlvs	28
	Aldebaran Tavrvs	29
	Aldebaren	32, 35
	Aldebora	33, 34
	BE	37, 42, 43, 44, 45
	Oc Tauri	38
	Ocul Tauri	39
	B. eye	40*sq*
	Bull^{s} Eye	41
	Bul Ey	43*sq*, 44*r*
Zeta	Cornu ♉ Aust:	24
Eta	Taigete Pleias	4
URSA MAJOR		
Alpha	Dubhe	1, 30, 32
	In Vrsa Maiore	4
	Cav, Vrse	5
	Il· Vr Ma	14
	Ursa	16*g*
Delta	El· C· V· M	14
	Elc	29
Epsilon	Principiu cauda urse maioris	10
	P· C· Vr· M	14
	Hume: Vrsa	16
	Alioth	18
	Prima cavde Vr Ma	19

URSA MAJOR *(cont.)*		Cat. No.
	Cavd. Vrs Ma	28
	Ilivm	29
Eta	Bencenaz	1
	eq^{9}	2, 3
	In Themone Prima	4
	Extre Cav Vr Ma	19
	Vlt. caud. Vrsae	24
	Vlt ca Vr	27
	Alkaid	32, 35
Theta	Svbpede	1
?	Maio	16*g*
	Vmb. Vrs. Ma	28
(5 star points)		
	vrsa ma	11
(7 star points)		
	Ursa maior	8, 9, 9*ua*
	Ursa major	25
URSA MINOR		
Alpha	E. C. ur. mi.	17
	Ext. Cau Vrs^{9} mn.	26
	Polaris	30
	Al	33, 34
Beta	B	33, 34
VELA	see Argo Navis	
VIRGO		
Alpha	Assimech	1
	Spica	2, 3, 4, 11, 30, 35
	Spica ♍	5, 9, 9*ua*, 10, 24, 26, 36
	Spica Virginis	8, 16*g*, 19, 21, 31
	Spica vir:	8*ua*
	Spica· Virg·	14
	Sp. v.	17
	Azimech	18
	Spic ♍	27
	Spica Virgo	29
	Spic Virg	38
Epsilon	Vindemiatrix	8
	Preuindemiat	16*g*

Comparison of Stars

Numbers listed on the backs of the retes of M-23 and M-24 (cat. nos. 8 and 9)

	M-23	M-24
1	Lanx boree cla:	Cauda ♑
2		Crus ♒
3		Cauda ceti
4	Crus aquarij	Venter ceti
5		Nares ceti
6		Oculus ♉
7		Orion: sinister hum:
8		Orio: sin: pes
9		Orionis dex: hum:
10		Canis maior
11	Caput draconis	Canicula
12		Cor ♌
13		Hÿdrae clara
14	Bootis sinister hu:	Crateris fundus
15	Arcturus	Spica ♍
16	Corona sept:	Cor ♏
17	Caput herculis	Ophiuc: ma: dex:
18	Lijra	Ophiuchi caput
19	Cauda cijgni	Caput herculis
20	Pectꝰ cassi	Caput draconis
21		Ophiuc: sin: manus
22	Meduse caput	Corona Septent:
23	Hircus	Lanx boree clarior
24	Ophiuc: cap:	Bootis sinist: hum:
25	Ophiuchi sinister ma:	Arcturus
26	Aquila	
27	Pegasi crus	
28	Pegasi humerus	
29		
30	Pegasi vmbi:	
31	Medi: cing: andro:	
32	Oculus tauri	
33		Cauda ♌
34		Dorsum ♌
35		Hircus
36	Cor leonis	Meduse caput
37	Cauda ♌	Pectꝰ cassi:
38		
39	Dorsū leonis	Pegasi vmbil:
40	Spica virginis	Crus pegasi
41		Pegasi humerus
42	Pegasi ala	Pegasi rictus
43		Cauda cÿgni
44		Aquila
45		Lÿra
46	Cauda capricorni	
47		
48		Cephei dex: hum:
49	Nares ceti	Andro: vmbilicꝰ
50		
51		Pers: latꝰ dex:
52	Orion: dexter hūerꝰ	
53		Ala pegasi
54	Orionis sinis: pes	
55		
56	Canis maior	Caput ♊ anter:
57		
58	Hijdre clara	Corui ala dextra
59	Crateris fundꝰ	
60	Venter ceti	
61		
62	Pegasi rictus	Ophiuchi genu dex:
63		
64		
65		Cingu: orioni media
66-88	[unused]	
89	Ophiuchi ma: dex:	

Note: On M-23, Scorpij cor, Ursa maior, Corui ala dextra, Canicula and Cauda ceti are not numbered, and Vindemiatrix is indicated by an asterisk. On M-24, only Ursa maior has no number.

Concordance

Cat. No.	Accession No.	Date
1	M-26	*c.* 1250
2	M-27	*c.* 1400
3	W-264	*c.* 1400
4	M-28	*c.* 1500?
5	W-272	1532
6	M-22	1540
7	M-20	*c.* 1550
8	M-23	1558
9	M-24	1564
10	M-25	*c.* 1600
11	M-21	1559
12	W-109	1567
13	M-45	before 1582
14	M-33A	1597
15	M-31	1598
16	M-33	*c.* 1620
17	M-42	early 17th cent.
18	M-34	1620
19	W-98	1584 and 1622
20	M-29	*c.* 1600
21	M-32	*c.* 1620
22	A-304A	*c.* 1628
23	A-304B	*c.* 1628
24	M-30	*c.* 1650
25	M-463	1682
26	M-465A	*c.* 1690
27	M-465B	*c.* 1700
28	DPW-51	19th cent.
29	A-111	modern
30	A-291	after 1908
31	W-97	1977
32	A-251	1985
33	W-250	1989
34	W-251	1989
35	W-253	*c.* 1989
36	A-108	*c.* 1550
37	DPW-10	*c.* 1650
38	W-256	1658
39	T-35	*c.* 1680
40	A-203	*c.* 1680
41	W-88	*c.* 1700
42	DPW-41	*c.* 1735
43	DPW-46	after 1752
44	W-70	after 1752
45	DPW-12	after 1752
46	A-275	1616
47	A-157	1963

Accession No.	Cat. No.
A-108	36
A-111	29
A-157	47
A-203	40
A-251	32
A-275	46
A-291	30
A-304A	22
A-304B	23
DPW-10	37
DPW-12	45
DPW-41	42
DPW-46	43
DPW-51	28
M-20	7
M-21	11
M-22	6
M-23	8
M-24	9
M-25	10
M-26	1
M-27	2
M-28	4
M-29	20
M-30	24
M-31	15
M-32	21
M-33	16
M-33A	14
M-34	18
M-42	17
M-45	13
M-463	25
M-465A	26
M-465B	27
T-35	39
W-70	44
W-88	41
W-97	31
W-98	19
W-109	12
W-250	33
W-251	34
W-253	35
W-256	38
W-264	3
W-272	5

Makers' Biographies

ARSENIUS, GUALTERUS
fl. 1554-1579
Louvain [Belgium — at that time a Spanish province]
Nephew of Gemma Frisius, for whom both Gualterus and his brother worked. He made a wide range of beautifully engraved instruments, including astrolabes, armillary spheres, astronomical rings, and radio astronomicos, all done to Gemma's designs.

BLAGRAVE, JOHN
c. 1558-1612
Reading, England
Designer of several unusual instruments, including two types of astrolabes, the Mathematical Jewel and the Uranical astrolabe, and the author of several books. In his youth he attended Oxford University.

BOS, JOHANNES
fl. 1591-1623
Antwerp; Rome
Maker of a series of astrolabes, all dated "24. Martii. 1597," that are smaller copies of the Adler instrument and are of rolled, rather than hammered, brass. He was the son of Jacob Bos, a cartographer.

DANFRIE, PHILIPPE
c. 1532-1608
Paris
Instrument maker, inventor of the Graphometer, *Controleur des Monnaies* to Henri III and IV, and author. In 1578 he engraved a set of copper plates for a paper astrolabe and, in 1584, corrected the changed calendar and reprinted the plates; these last plates were reprinted by Jehan Moreau (*q.v.*) in 1622.

FUSORIS, JOHANNE
c. 1365-1436
Paris
Renowned Parisian author and instrument maker, whose work included astrolabes, clocks, and equatoria. Exiled for consorting with the enemy — he had gone to England to try to get monies owed him by the Bishop of Norwich, a former ambassador from England to France — he was recalled to build a clock for the cathedral at Bourges, capital of the duchy of the Duc du Berry, brother of Charles V.

HABRECHT, ISAAC 2
1589-1633
Strasbourg, France
Maker of globes and planispheres and author of *Planiglobium coeleste, et terrestre* (1628). He was the son of Isaac Habrecht 1, the maker of the astronomical clock in Strasbourg.

HARRIS, DANIEL
fl. 1735-1775
London
Mathematical instrument maker. Apprenticed to Edmund Blow of the Joiners' Company in 1723, he was turned over to Thomas Cooke 1 in the same Company in 1725 and became free of the Company in 1735 as a mathematical instrument maker.

HARTMANN, GEORG
1489-1564
Nuremberg
Author, cleric, and one of the best and most prolific of the early German makers of astrolabes and sundials. He produced both brass and paper astrolabes.

HAYES, WALTER
fl. 1642-1692
London
Maker of mathematical instruments and quadrants, including unusual types of each. Apprenticed to John Allen 1 of the Grocers' Company in 1631, he became free of the Company in 1642. In 1667 he was admitted to the Clockmakers' Company, with eighteen other mathematical instrument makers.

LAUREN, CHRISTOPHOR
fl. *c.* 1598
Sens, France
Inventor of a type of astrolabe that was made by Ludovicus Martinot (*q.v.*).

MARTINOT, LUDOVICUS
fl. 1598-1631
Sens, France
Clockmaker and maker of a type of astrolabe invented by Christophor Lauren *(q.v.)*. The Martinot family produced many clockmakers, but we have found no reference to a Ludovicus Martinot in the various lists of clockmakers. He also made an astronomical volvelle in 1631.

MOREAU, JEHAN
fl. 1622-1628
Paris
Publisher and bookseller. In 1622, he reissued Philippe Danfrie's *(q.v.)* 1584 paper astrolabe, having added his name, address, and the new date to the plates.

PRUJEAN, JOHN
fl. 1670-1706
Oxford
Maker of brass and paper mathematical instruments and, from 1664, a nonacademic member of Oxford University. He was apprenticed to Thomas Alcock of the Clockmakers' Company in 1646 for eight years.

SCHRECKENFUCHS, LAURENTIUS
fl. *c.* 1567
Memmingen, Germany
Probably the son of Erasmus Oswald Schreckenfuchs of Memmingen, an author and an astrolabist.

SEVIN, PIERRE
fl. 1662-1688
Paris
Member of the *Corporation des Fondeurs, Ingénieur du Roi,* and maker of astrolabes, calendars, sundials (including handsome Butterfield types), and surveying instruments. He also divided instruments for Gosselin, and the Paris Observatory was one of his best clients.

SUTTON, HENRY
fl. 1648-1669
London
Maker of brass instruments and of engraved plates for paper astrolabes and astrolabe-quadrants. Apprenticed to Thomas Brown I of the Joiners' Company in 1636, he became free of the Company in 1648; he later took apprentices.

THOMPSON, ANTHONY
fl. 1638-1665
London
Maker of quadrants, dials, and other instruments, including some for Samuel Pepys. He took over the workshop of John Thompson, who was possibly his father.

VIBRANDI
c. 1900
Holland
Ghost-name invented for the maker of a fake astrolabe of the Bos-type (cat. no. 29).

WORGAN, JOHN
fl. 1682-1714
London
Maker of finely engraved instruments, including universal ring dials, other sundials, quadrants, and mathematical instruments, which he decorated with roses and tulips, and Master of Mechanics to King George I. Apprenticed to Nathaniel Anderton of the Grocers' Company in 1669, he became free of the Company in 1682.

ZABEUS, BERNARDINUS
fl. 1552-1559
Padua
Maker of astrolabes and sundials; M-21 (cat. no. 11) is signed and dated on the edge of the rim.

Bibliography

MANUSCRIPTS

Copenhagen, Det Kongelige Bibliotek, Cod. Hebr. 37 (illuminated Hebrew manuscript from Spain, *c.* 1350-99).

London, British Library, Add. MS 24189 (Czech manuscript of *Mandeville's Travels*).

London, British Library, MS Or. 10878 (Hebrew manuscript from Germany, *c.* 1400-50).

Oxford, Bodleian Library, MS Aubrey 10.

Oxford, Bodleian Library, MS Kennicott 1 (Hebrew-Spanish manuscript, 1472).

Paris, Bibliothèque de l'Arsenal, MS 1186 (13th-century manuscript).

ALMANACS

The American Ephemeris and Nautical Almanac. U.S. Government Printing Office. Washington, D.C. (Supt. of Documents, U.S. Government Printing Office, Washington, D.C. 20402.)

The Astronomical Ephemeris. H. M. Stationery Office. London. (H. M. Stationery Office, 49 High Holborn, London, WC1, England.)

The Nautical Almanac. U.S. Government Printing Office and H. M. Stationery Office. Washington, D.C./London.

BOOKS AND ARTICLES

Abbott (1937)	Abbott, Nadia. "Indian Astrolabe Makers." *Islamic Culture* 11 (Jan. 1937): 144-46.
Adler (1970)	*Adler Planetarium Guide Book.* Chicago, 1970.
Adler (1973)	*A Guide to the Adler Planetarium.* Chicago, 1973.
Adler (1977)	*The Adler Planetarium, 1976* (Annual Report, 1976). Chicago, 1977.
Adler (1980)	*Guide to the Adler Planetarium.* Chicago, 1980.
Adler (1989)	*The Adler Planetarium — 1988* (Annual Report, 1988). Chicago, 1989.
Adler (1992)	*The Adler Planetarium — 1991* (Annual Report, 1991). Chicago, 1992.
d'Ailly (*c.* 1480-83)	d'Ailly, Pierre. *Ymago mundi.* Louvain, [*c.* 1480-83].
Allen (1963)	Allen, Richard Hinckle. *Star Names: Their Lore and Meaning.* New York, 1963.
Annuli astronomici (1558)	*Annuli astronomici, instrumenti cum certissimi, tùm commodissimi, usus, ex variis authoribus, Petro Beausardo, Gemma Frisio, Ioãne Dryandro, Boneto Hebraeo, Burchardo Mythobio, Orontio Finaeo.* Paris, 1558.
Anthiaume and Sottas (1910)	Anthiaume, A., and Sottas, Jules. *L'astrolabe-quadrant du Musée des Antiquités de Rouen. Recherches sur les connaissances mathématiques, astronomiques et nautiques au moyen âge.* Paris, 1910.
Apianus (1540)	Apianus, Petrus. *Astronomicum caesareum.* Ingolstadt, 1540.
Apianus (1584)	Apianus, Petrus. *Cosmographia.* Antwerp, 1584.
Astrolabii...canones (1512)	*Astrolabii quo primi mobilis motus deprehenduntur canones.* Venice, 1512.

Augarde (1989) — Augarde, Jean-Dominique. "La fabrication des instruments scientifiques du XVIIIe siècle et la corporation des fondeurs." In *Studies in the History of Scientific Instruments,* edited by Christine Blondel *et al.,* 52-72. London/Paris, 1989.

Bartoli (1564 and 1614) — Bartoli, Cosimo. *Del modo di misurare le distantie, le superficie, i corpi, le piante, le provincie, le prospettive, & tutte le altre cose terrene, che possono occorrere a gli huomini.* Venice, 1564 and 1614.

Bayer (1603) — Bayer, Johann. *Uranometria.* Augsburg, 1603.

Behaim Catalogue (1992) — *Focus Behaim Globus.* Edited by Gerhard Bott. Catalogue of an exhibition at the Germanisches Nationalmuseum. 2 vols. Nuremberg, 1992.

Bennett (1987) — Bennett, James A. *The Divided Circle.* Oxford, 1987.

Bion (1702) — Bion, Nicolas. *L'usage des astrolabes, tant universels que particuliers.* Paris, 1702.

al-Bīrūnī (1934) — al-Bīrūnī, Abū Rayḥān. *Book of Instruction in the Elements of the Art of Astrology.* Translated by Robert Ramsay Wright. London, 1934.

al-Bīrūnī (1976) — al-Bīrūnī, Abū Rayḥān. *The Exhaustive Treatise on Shadows.* Translated by E. S. Kennedy. 2 vols. Aleppo, 1976.

Blagrave (1585) — Blagrave, John. *The Mathematical Jewel.* London, 1585.

Blundeville (1594) — Blundeville, Thomas. *His Exercises.* London, 1594.

Bourges (1994) — *L'horloge de Bourges.* Mécénat Technologique et Scientifique d'Electricité de France. Bourges, 1994.

Brahe (1598) — Brahe, Tycho. *Astronomiae instauratae mechanica.* Wandesburgi, 1598.

Brieux Catalogue (1983) — Brieux, Alain. *Instruments scientifiques.* Paris, 1983.

Brieux and Maddison (forthcoming) — Brieux, Alain, and Maddison, Francis R. *Répertoire des facteurs d'astrolabes et leurs ouvrages: Islam, plus Byzantine, Arménie, Géorgie et Inde hindoue.* Forthcoming.

Brussels Catalogue (1935) — *Cinq siècles d'art.* Catalogue of an exhibition, May 24-Oct. 13, 1935. Brussels, 1935.

Bryden (1993) — Bryden, David J. "Made in Oxford: John Prujean's 1701 Catalogue of Mathematical Instruments." *Oxoniensis* 58 (1993).

Bryden Catalogue (1988) — Bryden, David J. *Sundials and Related Instruments.* The Whipple Museum of the History of Science, Catalogue 6. Cambridge, England, 1988.

Burnett and Morrison-Low (1989) — Burnett, J. E., and Morrison-Low, A. D. *Vulgar and Mechanic: The Scientific Instrument Trade in Ireland, 1650-1921.* Dublin, 1989.

Capo (1994) — Capo, Bernard. *L'horloge astronomique de Bourges — histoire d'une réhabilitation.* Bourges, 1994.

Cardinal (1986) — Cardinal, Catherine. "Horloges de table astrolabiques françaises du XVIe siècle." *Astrolabica* 4 (1986): 3-20.

Chapiro *et al.* Catalogue (1989) — Chapiro, Adolphe; Meslin-Perrier, Chantal; and Turner, Anthony. *Catalogue de l'horlogerie et des instruments de précision du début du XVIe au milieu du XVIIe siècle.* Musée National de la Renaissance — Chateau d'Ecouen. Paris, 1989.

Chardin (1735) — Chardin, John. *Voyages du Chevalier Chardin, en Perse, et autres lieux de l'Orient.* New ed. 4 vols. Amsterdam, 1735.

Chaucer (1391) — Chaucer, Geoffrey. *Treatise on the Astrolabe.* 1391. Revised/modernized text based on the Rawlinson MS printed in Gunther (1929a).

Chaucer (1872) — Chaucer, Geoffrey. *A Treatise on the Astrolabe addressed to his son Lowys, A.D. 1391.* Edited by Walter W. Skeat. London, 1872.

Chaucer (1987) — Chaucer, Geoffrey. *The Riverside Chaucer.* 3rd ed. Edited by Larry D. Benson. Boston, 1987.

Christie's New York (1985) — Christie's New York. *Fine Scientific Instruments, Clocks, Watches and Related Books.* Catalogue of sale on Oct. 31, 1985. New York, 1985. (Ref. 100-101, lot 334.)

Christie's New York (1988) — Christie's New York. *Gold and Silver of the Atocha and Margarita.* Catalogue of sale on June 14-15, 1988. New York, 1988. (Ref. 62-63, lot 20.)

Circa 1492 Catalogue (1991) — *Circa 1492: Art in the Age of Exploration.* Edited by Jay A. Levenson. Catalogue of an exhibition at the National Gallery of Art, Washington, D.C., Oct. 12, 1991-Jan. 12, 1992. New Haven/London, 1991.

Clavius (1593) — Clavius, Christoph. *Astrolabium.* Rome, 1593.

Cohen (1947) — Cohen, A., ed. *The Soncino Chumash: The Five Books of Moses with Haphtaroth.* London, 1947.

Collins (1658) — Collins, John. *The Sector on a Quadrant or A Treatise containing the Description and use of three several Quadrants.* London, 1658.

Combe *et al.* (1931-54) — Combe, Etienne; Sauvaget, Jean; and Wiet, Gaston. *Répertoire chronologique d'epigraphie arabe.* Vol. 10. Cairo, 1931-54.

Copp (1525) — Copp, Johann. *Erklaerung unnd Gründtliche underweysung, alles nutzes, so in dem Edlen Instrument, Astrolabiū genañt.* Augsburg, 1525.

Crawforth (1987) — Crawforth, Michael. "Instrument Makers in the London Guilds." *Annals of Science* 44 (1987): 319-77.

Curator (1995) — *Curator: The Museum Journal* 38, no. 3 (Sept. 1995): cover.

Danti (1569) — Danti, Egnatio. *Trattato dell'uso et della fabbrica dell'astrolabio.* Florence, 1569.

Danti (1578) — Danti, Egnatio. *Dell'uso et fabbrica dell'astrolabio et del planisferio.* Florence, 1578.

Debeauvais and Marche (in preparation) — Debeauvais, Francis, and Marche, Robert. *Etude de planiglobe céleste d'Isaac Habrecht II (1589-1633) réf. A-304A de l'Adler Planetarium de Chicago, 1995.* In preparation.

Destombes (1962) — Destombes, Marcel. "Un astrolabe carolingien et l'origine de nos chiffres arabes." *Archives internationales d'histoire des sciences* 58-59 (1962): 3-45.

Dor-Nev (1992) — Dor-Nev, Zvi. *Columbus and the Age of Discovery.* New York, 1992.

Drake Catalogue (1979) — *Sir Francis Drake.* Catalogue of an exhibition at the Bancroft Library, University of California, Berkeley, June 14-Oct. 6, 1979. Berkeley, 1979.

Dreier (1979) — Dreier, Franz Adrian. *Winkelmessinstrumente vom 16, bis zum Frühen 19, Jahrhundert.* Berlin, 1979.

Dryander (1537) — Dryander, Johann. *Annulorum trium diversi generis instrumentorum astronomicorum, componendi ratio atq; usus, cum quibusdam aliis lectu iucundissimis.* Marburg, 1537.

Edwards and Signell Catalogue (1982) — Edwards, Holly, and Signell, Karl. *Patterns and Precision: The Arts and Sciences of Islam.* Catalogue of an exhibition. Washington, D.C., 1982.

Encyclopaedia Judaica (1973) — *Encyclopaedia Judaica.* Edited by Cecil Roth. Jerusalem/New York, 1973.

Engelmann Catalogue (1924) — Engelmann, Max. *Sammlung Mensing. Altwissenschaftliche Instrument Katalog.* 2 vols. Amsterdam, 1924.

Faustus (*c.* 1680) — *First Part of Dr. Faustus, Abreviated and brought into verse.* London, [*c.* 1680].

Fernandez Villars (1976) — Fernandez Villars, Miguel Angel. *Sobre el astrolábio firmado por G. Frisius y G. Arsenius.* Museo Nacional de Historia: Castillo de Chapultepec. Mexico City, 1976.

Fine (1534) — Fine, Oronce. *Quadrans astrolabicus.* Paris, 1534.

Fine (1542) — Fine, Oronce. *De mundi sphaera.* Paris, 1542.

Fischer *et al.* (1988) — Fischer, K. A. F.; Kunitzsch, P.; and Langermann, Y. T. "The Hebrew Astronomical Codex MS Sassoon 823." *Jewish Quarterly Review* 78 (1988): 253-92.

Focard (1546 and 1555) — Focard, Jacques. *Paraphrase de l'astrolabe.* Lyons, 1546. Revised by Jacques Bassentin, Lyons, 1555.

Fox (1932) — Fox, Philip. "The Adler Planetarium and Astronomical Museum of Chicago." *Popular Astronomy* 40 (March 1932): 125-55, 321-51, 532-49, 613-22.

Fox (1933) — Fox, Philip. *The Adler Planetarium and Astronomical Museum: An Account of the Optical Planetarium and a Brief Guide to the Museum.* Chicago, 1933. Reprinted 1942.

F. P. (1669) — F[rancis] P[otter]. *A Description and Use of A large Quadrant contrived and made by Henry Sutton, with The Description and use of a geodaetical Scheme.* London, 1669.

Gabb (1937) — Gabb, George H. "The Astrological Astrolabe of Queen Elizabeth." *Archaeologia* 86 (1937): 101-3.

Gallucci (1597) — Gallucci, G. Paolo. *Della fabrica et uso di diversi stromenti de astronomia.* Venice, 1597.

Gandz (1927) — Gandz, Solomon. "The Astrolabe in Jewish Literature." *Hebrew Union College Annual* 4 (1927): 469-86.

García Franco (1945) — García Franco, Salvador. *Catálogo crítico de astrolabios existentes en España.* Madrid, 1945.

Gemma Frisius (1533) — Gemma Frisius. *Libellus de locorum describendorum ratione.* Antwerp, 1533.

Gemma Frisius (1556 and 1583) — Gemma Frisius. *De astrolabo catholico liber.* Antwerp, 1556 and 1583. Reissued as part of Apianus (1584).

Gent (1994) — Gent, R. H. van. *The Portable Universe: Two Astrolabes of the Museum Boerhaave.* Leiden, 1994.

Gibbs *et al.* (1973) — Gibbs, Sharon L.; Henderson, Janice A.; and Price, Derek J. de Solla. *A Computerized Checklist of Astrolabes.* New Haven, 1973. (Computer printout of ICA checklist.)

Gibbs with Saliba (1984) — Gibbs, Sharon L., with Saliba, George. *Planispheric Astrolabes from the National Museum of American History.* Smithsonian Studies in History and Technology, no. 45. Washington, D.C., 1984.

Gingerich (1987) — Gingerich, Owen. "Zoomorphic Astrolabes and the Introduction of Arabic Star Names into Europe." In *From Deferent to Equant: A Volume of Studies in the History of Science in the Ancient and Medieval Near East in Honor of E. S. Kennedy*, edited by David A. King and George Saliba, 89-104. Annals of the New York Academy of Sciences, vol. 500. New York, 1987.

Gingerich *et al.* (1972) — Gingerich, Owen; King, David; and Saliba, George. "The 'Abd al-A'immah Astrolabe Forgeries." *Journal of the History of Astronomy* 3 (1972): 188-98. Reprinted in King (1987a), paper no. VI.

Goldstein (1976) — Goldstein, Bernard R. "The Hebrew Astrolabe in the Adler Planetarium." *Journal of Near Eastern Studies* 35 (1976): 251-60. Reprinted in Goldstein (1985b), paper no. XVIII.

Goldstein (1977) — Goldstein, Bernard R. "Levi ben Gerson: On Instrumental Errors and the Transversal Scale." *Journal for the History of Astronomy* 8 (1977): 102-12.

Goldstein (1985a) — Goldstein, Bernard R. *The Astronomy of Levi Ben Gerson (1288-1344).* New York/Berlin, 1985.

Goldstein (1985b) — Goldstein, Bernard R. *Theory and Observation in Ancient and Medieval Astronomy.* Variorum Reprints. London, 1985.

Goldstein (1985c) — Goldstein, Bernard R. "Star Lists in Hebrew." *Centaurus* 28 (1985): 185-208.

Goldstein and Chabás (1996) — Goldstein, Bernard R., and Chabás, J. "Ibn al-Kammād's Star List." *Centaurus* 38 (1996): 317-34.

Goldstein and Saliba (1983) — Goldstein, Bernard R., and Saliba, George. "A Hispano-Arabic Astrolabe with Hebrew Star Names." *Annali dell'Istituto e Museo di Storia della Scienza di Firenze* 8 (1983): 19-28. Reprinted in Goldstein (1985b), paper no. XIX.

Grant (1974) — Grant, Edward, ed. *A Source Book in Medieval Science.* Cambridge, Mass., 1974.

Greene (1977) — Greene, Norman, publ. *Chaucer on the Astrolabe, with original illustrations.* Revised from the 1931 Oxford edition by R. T. Gunther. Berkeley, 1977.

Guiffrey (1894-96) — Guiffrey, Jules. *Inventaires de Jean, duc de Berry (1401-1416).* 2 vols. Paris, 1894-96.

Gunter (1624) — Gunter, Edmund. *The Description and Use of the Sector, the Crosse-staffe and other instruments.* London, 1624.

Gunter (1653) — Gunter, Edmund. *The Works of Edmund Gunter containing the description and use of the Sector, Cross-Staff and other instruments/whereunto is now added the further use of the quadrant by Samuel Foster.* 3rd ed., corrected and amended by Henry Bond. London, 1653.

Gunter (1662) — Gunter, Edmund. *The Works of Edmund Gunter containing the description and use of his Sector, Cross-Staff, Bow, Quadrant and Other Instruments.* London, 1662.

Gunther (1929a) — Gunther, Robert T. *Chaucer and Messahalla on the Astrolabe.* Early Science in Oxford, vol. 5. Oxford, 1929.

Gunther (1929b) — Gunther, Robert T. "The Uranical Astrolabe and other Inventions of John Blagrave of Reading." *Archaeologia* 2nd series XXIX (1929): 55-72.

Gunther (1932) — Gunther, Robert T. *The Astrolabes of the World.* 2 vols. Oxford, 1932.

Gunther (1936) — Gunther, Robert T. "The Newly Found Astrolabe of Queen Elizabeth." *The Illustrated London News* (Oct. 24, 1936): 738-39.

Gunther (1937a) — Gunther, Robert T. *Early Science at Cambridge.* Oxford, 1937.

Gunther (1937b) — Gunther, Robert T. "The Astrolabe of Queen Elizabeth." *Archaeologia* 86 (1937): 65-72.

Habrecht (1628) — Habrecht, Isaac 2. *Planiglobium coeleste, et terrestre.* Strasbourg, 1628.

Habrecht (1666) — Habrecht, Isaac 2. *Planiglobium coeleste ac terrestre argentorati quondam, nunc operâ Johannis Christophori Sturmii.* Nuremberg, 1666.

Harington (1596) — Harington, Sir John. *A New Discourse of a Stale Subject, Called the Metamorphosis of Ajax.* London, 1596.

Hart *et al.* (1979) — Hart, James; Parker, Earl; Michel, H. V.; Asaro, F.; and Norberg, A. L. *The Plate of Brass Re-examined: A Supplement.* Berkeley, 1979.

Hartner (1939) — Hartner, Willy. "The Principle and Use of the Astrolabe." In *A Survey of Persian Art from Prehistoric Times to the Present*, edited by Arthur Upham Pope, 3: 2530-54. London/New York, 1939. Reprinted in Hartner (1968/1984), 1: 287-311.

Hartner (1950) — Hartner, Willy. "The Astronomical Instruments of Cha-ma-lu-ting, Their Identification, and Their Relations to the Instruments of the Observatory of Marāgha." *Isis* 41 (1950): 184-94.

Hartner (1960) — Hartner, Willy. "Asṭurlāb." In *Encyclopedia of Islam,* new ed., 1: 722-28. [Leiden], 1960. Reprinted in Hartner (1968/1984), 1: 312-18.

Hartner (1968/1984) — Hartner, Willy. *Oriens-Occidens.* 2 vols. Hildesheim, 1968/1984.

Haskins (1927) — Haskins, Charles Homer. *Studies in the History of Medieval Science.* 2nd ed. Cambridge, Mass., 1927.

Hayward Catalogue (1975) — *The Secular Spirit: Life and Art at the End of the Middle Ages.* Edited by Jane Hayward. Catalogue of an exhibition at The Cloisters. New York, 1975.

Heilbronner Catalogue (1922) — *Instruments de mathématiques.* Catalogue of the Heilbronner sale at Hôtel Drouot, March 8, 1922. Paris, 1922.

Henrion (1620) — Henrion, Denis. *Briefve explication de l'usage de l'astrolabe.* Paris, 1620.

Henrion (1621) — Henrion, Denis. *Collection, ou recueil de divers traictez mathematiques.* Paris, 1621.

Holbrook *et al.* (1992) — Holbrook, Mary; Anderson, Robert; and Bryden, David. *Science Preserved.* London, 1992.

ICA — See Gibbs *et al.* (1973) (International Checklist of Astrolabes).

Impey and MacGregor (1985) — Impey, Oliver, and MacGregor, Arthur, eds. *The Origins of Museums: The Cabinet of Curiosities in Sixteenth- and Seventeenth-Century Europe.* Oxford, 1985.

Jacquinot (1559) — Jacquinot, Dominique. *L'usage de l'astrolabe....plus est adjousté une amplification de l'usage de l'astrolabe, par Jacques Bassentin Escossois.* 2nd ed. Paris, 1559.

Jacquinot (1625) — Jacquinot, Dominique. *L'usage de l'un et l'autre astrolabe particulier et universel.* Paris, 1625.

Jenkin (1925) — Jenkin, C. F. *The Astrolabe, Its Construction and Use.* Oxford, 1925.

Josten (1954) — Josten, C. H. *Scientific Instruments (13th-19th Centuries): The Collection of J. A. Billmeir.* Oxford, 1954.

Kennedy and Kennedy (1987) — Kennedy, E. H., and Kennedy, M. H. *Geographical Coordinates of Localities from Islamic Sources.* Frankfurt am Main, 1987.

Kepler (1627) — Kepler, Johannes. *Tabulae Rudolphinae.* Ulm, 1627.

Kiely (1979) — Kiely, Edmond R. *Surveying Instruments: Their History.* Columbus, Ohio, 1979.

King (1972) — See Gingerich *et al.* (1972).

King (1974) — King, David A. "An Analog Computer for Solving Problems of Spherical Astronomy: The Shakkāzīya Quadrant of Jamāl al-dīn al-Māridīnī." *Archives internationales d'histoire des sciences* 24 (1974): 219-42. Reprinted in King (1987a), paper no. X.

King (1979a) — King, David A. "The Astronomical Instruments of Ibn al Sarrāj: A Brief Survey." Paper presented at the Second International Symposium for the History of Arabic Science, Aleppo, 1979. Reprinted in King (1987a), paper no. IX.

King (1979b) — King, David A. "Kibla." In *Encyclopedia of Islam*, new ed., 5: 83-88. 1979.

King (1979c) — King, David A. "On the Early History of the Universal Astrolabe in Islamic Astronomy and the Origin of the term 'Shakkāzīya' in Medieval Scientific Arabic." *Journal for the History of Arabic Science* 3 (1979): 244-57. Reprinted in King (1987a), paper no. VII.

King (1981) — King, David A. "The Origin of the Astrolabe According to Medieval Islamic Sources." *Journal for the History of Arabic Science* 5 (1981): 43-83. Reprinted in King (1987a), paper no. III.

King (1984) — King, David A. *Astronomy for Landlubbers and Navigators: The Case of the Islamic Middle Ages.* Centro de Estudos de História e Cartografia Antiga, no. 164. Lisbon, 1984.

King (1987a) — King, David A. *Islamic Astronomical Instruments.* Variorum Reprints. London, 1987.

King (1987b) — King, David A. "The Astrolabe of 'Alī al-Wadā'ī." In King (1987a), paper no. VIII.

King (1988) — King, David A. "Universal Solutions to Problems of Spherical Astronomy from Mamluk Egypt and Syria." In *A Way Prepared: Essays on Islamic Culture in Honor of Richard Bayly Winder,* edited by F. Kazeni and R. D. McChesney, 153-84. New York, 1988. Reprinted in King (1993), paper no. VII.

King (1991) — King, David A. "Strumentazione astronomica nel mondo medievale islamico." In *Storia della scienza: Gli strumenti*, 154-89. Turin, 1991.

King (1992) — King, David A. "Die Astrolabiensammlung des Germanischen Nationalmuseums." Translated by Kurt Maier. In Behaim Catalogue (1992), 1: 101-14, 2: 568-602, 640-43.

King (1993) — King, David A. *Astronomy in the Service of Islam.* Variorum Reprints. Aldershot, England, 1993.

King (1994) — King, David A. "Astronomical Instruments between East and West." In *Kommunikation Zwischen Orient und Okzident Alltag und Sachkultur,* 143-98. Österreichische Akademie der Wissenschaften Philosophisch-Historische Klasse Sitzungsberichte, vol. 619. Vienna, 1994.

Klemm (1990) — Klemm, Hans G. "Georg Hartmann aus Eggolsheim (1489-1564): Leben und Werk eines fränkischen Mathematikers und Ingenieurs." In *Wissenschaftliche und künstlerische Beiträge Ehrenbürg-Gymnasium Forchheim,* 8: 29, 80-81. Forchheim, 1990.

Köbel (1535) — Köbel, Jacob. *Astrolabii declaratio, eiusdemque usus mire jucundus, non modo astrologis, medicis, geographis, caeterisque literarum cultoribus multum utilis ac necessarius; verum etiam mechanicis quibusdam opificib. non parum commodus.* Mainz, 1535.

Krása (1983) — Krása, Josef, ed. *The Travels of Sir John Mandeville: A Manuscript in the British Library.* Translated by Peter Kussi. New York, 1983.

Kunitzsch (1959) — Kunitzsch, Paul. *Arabische Sternnamen in Europa.* Wiesbaden, 1959.

Kunitzsch (1961) — Kunitzsch, Paul. *Untersuchungen zur Sternnomenklatur der Araber.* Wiesbaden, 1961.

Kunitzsch (1981a) — Kunitzsch, Paul. "Observations on the Arabic Reception of the Astrolabe." *Archives internationales d'histoire des sciences* 31 (1981): 243-52.

Kunitzsch (1981b) — Kunitzsch, Paul. "On the Authenticity of the Treatise on the Composition and Use of the Astrolabe Ascribed to Messahalla." *Archives internationales d'histoire des sciences* 31 (1981): 42-62.

Labarte (1879) — Labarte, Jules. *Inventaire du mobilier de Charles V, roi de France.* Paris, 1879.

Lamprey (1997) — Lamprey, John P. "An Examination of Two Groups of Georg Hartmann Sixteenth-Century Astrolabes and the Tables Used in Their Manufacture." *Annals of Science* 54 (1997): 111-42.

Lansberg (1635) — Lansberg, Philipp. *Quadrantem tum astronomicum tum geometricum nec non in astrolabium.* Rome, 1635.

Lattes (1493) — Lattes, Bonet de. *Annulus astronomicus.* Rome, 1493.

Levi-Donati (1993) — Levi-Donati, G. R. "Uno strumento ritrovato: l'astrolabio perugino dell'anno 1498." *Bolletino della Deputazione di Storia Patria per l'Umbria* 90 (1993): 80-107.

Leybourn (1672) — Leybourn, William. *Panorganon: A Universall Instrument.* London, 1672.

Leybourn (1731) — Leybourn, William. *The Description and Use of a Portable Instrument, vulgarly known by the name of Gunter's Quadrant.* 3rd ed., with addition by Charles Leadbetter. London, 1731.

Lindberg (1978) — Lindberg, David C. "The Transmission of Greek and Arabic Learning to the West." In *Science in the Middle Ages,* edited by David C. Lindberg, 52-90. Chicago, 1978.

Longnon and Cazelles (1969) — Longnon, Jean, and Cazelles, Raymond. *The Très Riches Heures of Jean, Duke of Berry.* New York, 1969.

Maddison (1957) — Maddison, Francis R. *A Supplement to a Catalogue of Scientific Instruments in the Collection of J. A. Billmeir, Esq. C.B.E.* Oxford, 1957.

Maddison (1962) — Maddison, Francis R. "A 15th Century Islamic Spherical Astrolabe." *Physis* 4 (1962): 101-9.

Maddison (1966) — Maddison, Francis R. *Hugo Helt and the Rojas Astrolabe Projection.* Agrupamento de Estudos de Cartografia Antiga, no. 12. Coimbra, 1966. Appendix.

Maddison (1969) — Maddison, Francis R. *Medieval Scientific Instruments and the Development of Navigational Instruments in the XVth and XVIth Centuries.* Agrupamento de Estudos de Cartografia Antiga, no. 30. Coimbra, 1969.

Maddison (1991) — Maddison, Francis R. "Measuring and Mapping." In *Circa 1492* Catalogue (1991), 224-25, cat. entry 123.

Maddison and Turner Catalogue (1976) — Maddison, Francis R., and Turner, A. J. "Science and Technology in Islam." Catalogue of an exhibition at the Science Museum, London, April-Aug. 1976, in association with the Festival of Islam. Not published; privately circulated in a xerographic edition of 50 copies.

Mann *et al.* Catalogue (1992) — Mann, V. B., *et al. Convivencia: Jews, Muslims and Christians in Medieval Spain.* Catalogue of an exhibition at The Jewish Museum, Sept. 20-Dec. 20, 1992. New York, 1992.

Maurice and Mayr (1980) — Maurice, Klaus, and Mayr, Otto, eds. *The Clockwork Universe: German Clocks and Automata, 1550-1650.* New York, 1980.

Mayer (1956) — Mayer, Leo Ary. *Islamic Astrolabists and Their Works.* Geneva, 1956.

Metzger and Metzger (1982) — Metzger, Thérèse, and Metzger, Mendel. *Jewish Life in the Middle Ages: Illuminated Hebrew Manuscripts of the Thirteenth to the Sixteenth Centuries.* New York, 1982.

Michel (1935) — Michel, Henri. "L'art des instruments de mathématiques en Belgique au XVIe." *Bulletin Société Royale d'Archéologie de Bruxelles,* no. 2 (March-April 1935): 8, fig. 6.

Michel (1947) — Michel, Henri. *Traité de l'astrolabe.* Paris, 1947.

Millàs-Vallicrosa (1963) — Millàs-Vallicrosa, J. M. "Translations of Oriental Scientific Works (to the End of the Thirteenth Century)." In *The Evolution of Science,* edited by Guy S. Metraux and François Crouzet, 128-67. New York, 1963.

Milo *et al.* Catalogue (1955) — Milo, T. H., *et al. Hollands Glorie: de maritieme geschiedenis van Nederland.* Catalogue of an exhibition at Museum Het Prinsenhof, May-Aug. 1955. Delft, 1955.

Miniati Catalogue (1991) — Miniati, Mara. *Catalogo di Museo de Storia della Scienza.* Florence, 1991.

Mollan (1990) — Mollan, Charles. *Irish Interim Inventory.* Dublin, 1990.

Moreau (1625) — Moreau, Jean. *L'usage de l'un et l'autre astrolabe particulier et universel.* Paris, 1625.

Morley (1856) — Morley, William H. *Description of a Planispheric Astrolabe Constructed for Shah Sultan Husain Safawi King of Persia and Now Preserved in the British Museum.* London, 1856.

Morrison (1994) — Morrison, James E. "The Electronic Astrolabe." *Interdisciplinary Science Reviews* 19 (1994): 55-69.

Multhauf (1971) — Multhauf, Robert P. *Laurits Christian Eichner — Craftsman, 1894-1967.* Washington, D.C., 1971.

Murdoch (1984) — Murdoch, John E. *Albums of Science: Antiquity and the Middle Ages.* New York, 1984.

Nadvi (1935) — Nadvi, Sayyid Sulayman. "Some Indian Astrolabe Makers." *Islamic Culture* 9 (Oct. 1935): 621-31.

Nadvi (1937) — Nadvi, Sayyid Sulayman. "Indian Astrolabe Makers." *Islamic Culture* 11 (Oct. 1937): 537-39.

National Maritime Museum Astrolabes (1976) — [Howse, Derek, *et al.*] National Maritime Museum. *The Planispheric Astrolabe.* Greenwich, 1976.

National Maritime Museum Catalogue (1970) — *National Maritime Museum Inventory.* Vol. 2. London, 1970.

Needham (1954-88) — Needham, Joseph. *Science and Civilisation in China.* 6 vols. Cambridge, England, 1954-88.

Neugebauer (1949) — Neugebauer, Otto. "The Early History of the Astrolabe: Studies in Ancient Astronomy IX." *Isis* 40 (1949): 240-56.

Neugebauer (1975) — Neugebauer, Otto. *A History of Ancient Mathematical Astronomy.* 3 vols. New York, 1975.

North (1966) — North, John D. "Werner, Apian, Blagrave and the Meteoroscope." *British Journal for the History of Science* 3, part 1, no. 9 (June 1966): 57-65, plate.

North (1969) — North, John D. "Kalenderes Enlumyned Ben They: Some Astronomical Themes in Chaucer." *The Review of English Studies* n.s., 20 (1969): 129-54, 257-83, 418-44.

North (1974) — North, John D. "The Astrolabe." *Scientific American* 230 (Jan. 1974): 96-106. Reprinted in North (1989), 211-20.

North (1981) — North, John D. "Astrolabes and the Hour-Line Ritual." *Journal for the History of Arabic Science* 5 (1981): 113-14.

North (1988) — North, John D. *Chaucer's Universe.* Oxford, 1988.

North (1989) — North, John D. *Stars, Minds and Fate: Essays in Ancient and Medieval Cosmology.* London, 1989.

Osley (1969) — Osley, A. S. *Mercator: A Monograph on the Lettering of Maps.* London, 1969.

Oxford Universal Dictionary (1955) — *Oxford Universal Dictionary.* Edited by C. T. Onions. 3rd ed. Oxford, 1955.

Palmer (1658) — Palmer, John. *The Mathematical Jewell.* London, 1658.

Paris Universal Exhibition Catalogue (1900) — *L'exposition universelle internationale, Musée Retrospectif, classe 15.* Paris, 1900.

Pedersen (1978) — Pedersen, Olaf. "Astronomy." In *Science in the Middle Ages,* edited by David C. Lindberg, 303-37. Chicago, 1978.

Phillip, Son, & Neale Catalogue (1971) — Phillip, Son, & Neale. *Sale Catalogue.* Catalogue of sale on Feb. 16, 1971. London, 1971. (Ref. lot 135.)

Philopon (1981) — Philopon, Jean. *Traité de l'astrolabe.* Edited and translated by A. P. Segonds. *Astrolabica* 2. Paris, 1981.

Pingree (1978a) — Pingree, David. "History of Mathematical Astronomy in India." In *Dictionary of Scientific Biography,* 15: 533-633. New York, 1978.

Pingree (1978b) — Pingree, David. "Islamic Astronomy in Sanskrit." *Journal for the History of Arabic Science* 2 (1978): 315-30.

Poblacion (1527) — Poblacion, Joannis Martini [pseud. of Juan Martínez Siliceo]. *De usu astrolabii compendium.* [Paris], 1527.

Pogo (1934) — Pogo, Alexander. "Gemma Frisius, His Method of Determining Differences of Longitude by Transporting Timepieces (1530), and His Treatise on Triangulation (1533). With…a Facsimile Reproduction (No. XVI) of Gemma's *Libellus de locorum describendorum ratione,* Antwerp, 1533." *Isis* 22 (1934): 469-504.

Poulle (1954) — Poulle, Emmanuel. "L'astrolabe médiévale d'après les manuscrits de la Bibliothèque Nationale." *Bibliothèque de l'École de Chartes* 112 (1954): 81-103.

Poulle (1955) — Poulle, Emmanuel. "La fabrication des astrolabes au moyen âge." *Techniques et civilizations* 4 (1955): 117-28.

Poulle (1963) — Poulle, Emmanuel. *Un constructeur d'instruments astronomiques au XVe siècle: Jean Fusoris.* Paris, 1963.

Poulle (1967) — Poulle Emmanuel. *Le navire et l'économie maritime du XVe au XVIIIe siècle.* Paris, 1967.

Poulle (1980a) — Poulle, Emmanuel. *Les instruments de la théorie des planètes selon Ptolémée: Équatoires et horlogerie planétaire du XIIIe au XVIe siècle.* 2 vols. Geneva, 1980.

Poulle (1980b) — Poulle, Emmanuel. *Walcher de Malvern et son astrolabe (1092).* Centro de Estudos de Cartografia Antiga, Secção de Coimbra, no. 132. Coimbra, 1980.

Price (1955) — Price, Derek J. de Solla. "An International Checklist of Astrolabes." *Archives internationales d'histoire des sciences,* nos. 32 and 33 (1955): 243-63, 363-81.

Price (1956) Price, Derek J. de Solla. "Fake Antique Scientific Instruments." In *Actes du VIIIe congrès international d'histoire des sciences* (1956), 1: 380-99. Collection de trauvaux de l'Académie Internationale d'Histoire des Sciences, no. 9. Florence, 1958.

Price (1957) Price, Derek J. de Solla. "Precision Instruments to 1500." In *A History of Technology,* edited by Charles Singer *et al.*, 3: 582-619. Oxford, 1957.

Puig (1986) Puig, Roser. *Al-Šakkāzıyya Ibn al-Naqqāš al-Zarqālluh.* Barcelona, 1986.

Ræder *et al.* (1946) Ræder, Hans; Strömgren, Elis; and Strömgren, Bengt, eds. and trans. *Tycho Brahe's Description of His Instruments and Scientific Work.* Copenhagen, 1946.

Rees (1819) Rees, Abraham. *The Cyclopedia; or, Universal Dictionary of Arts, Sciences, and Literature.* Philadelphia, 1819.

Regnartius (1610) Regnartius, Valerianus. *Astrolabiorum seu utriusque planisphaerii universalis et particularis usus.* Rome, 1610.

Renaud (1932) Renaud, H. P. J. "Additions et corrections à Suter 'Die Mathematiker und Astronomen der Araber.'" *Isis* 18 (1932): 166-83.

Renaud (1942) Renaud, H. P. J. "Quelques constructeurs d'astrolabes en occident musulman." *Isis* 34 (1942): 20-23.

Rico y Sinobas (1863-68) Rico y Sinobas, Manuel, ed. *Libros del saber de astronomia del Rey D. Alfonso X de Castilla.* 5 vols. Madrid, 1863-68.

Ritter (1613) Ritter, Franz. *Astrolabium.* Nuremberg, [1613].

Rohde (1923) Rohde, Alfred. *Die Geschichte des Wissenschaftlichen Instrumente.* Leipzig, 1923.

de Rojas (1550) de Rojas, Juan. *Commentariorum in astrolabium quod planisphærium vocant, libri sex.* Paris, 1550.

de Rojas (1551) See de Rojas (1550).

Rooseboom (1950) Rooseboom, Maria. "Bijdrage tot de Geschiedenis der Instrumentmakerskunst in de Noordelijke Nederlanden tot Omstreeks 1840." *Mededeling No. 74 uit het Rijksmuseum voor de Geschiedenis de Natuurweten-schappen te Leiden* (1950).

Roussel Catalogue (1911) *Roussel Sale.* Catalogue of sale at Hôtel Drouot. Paris, 1911. (Ref. 41, lot 196.)

Sacrobosco (1485) Sacrobosco, Joannes de. *Spherae Mundi.* Augsburg, 1485.

Sacrobosco (1500) Sacrobosco, Joannes de. *Textus de sphera.* Paris, 1500.

Sacrobosco (1579) Sacrobosco, Joannes de. *La sfera di messer Giovanni Sacrobosco tradotta, emendata & distinta in capitoli da Pieruincenzo Dante de' Rinaldi con molte, et vtili annotazioni del medesimo.* Florence, 1579.

Saliba (1977) Saliba, George. "The Buffalo Astrolabe of Muhammad Khalil." *Al-Abhāth, Quarterly Journal of the American University of Beirut* 26 (1977): 11-18.

Saunders (1971) Saunders, Harold N. *The Astrolabe Kit.* Bude, England, 1971.

Saunders (1984) Saunders, Harold N. *All the Astrolabes.* Oxford, 1984.

Savage-Smith (1985) Savage-Smith, Emilie. *Islamicate Celestial Globes: Their History, Construction, and Use.* Smithsonian Studies in History and Technology, no. 46. Washington, D.C., 1985.

Savage-Smith (1990) Savage-Smith, Emilie. "The Classification of Islamic Celestial Globes in the Light of Recent Evidence." *Der Globusfreund* 38/39 (Nov. 1990): 23-29.

Savage-Smith (1992) Savage-Smith, Emilie. "Celestial Mapping." In *The History of Cartography,* edited by David Woodward. Vol. 2, book 1, *Cartography in the Traditional Islamic and South Asian Societies*, 12-70, plates 1-2. Chicago, 1992.

Schechner Genuth (1997) Schechner Genuth, Sara. *Comets, Popular Culture, and the Birth of Modern Cosmology.* Princeton, 1997.

Schedel (1493) Schedel, Hartmann. *Büch der Cronicken.* Nuremberg, 1493.

Schroeder (1956) — Schroeder, Wolfgang. *Practical Astronomy: A New Approach to an Old Science*. London, 1956.

Seymour (1968) — Seymour, M. C., ed. *Mandeville's Travels*. London, 1968.

Sharma (1984) — Sharma, Virendra Nath. "The Great Astrolabe of Jaipur and Its Sister Unit." *Archaeoastronomy*, no. 7, *Supplement to the Journal for the History of Astronomy* 15 (1984): S126-28.

Siraisi (1990) — Siraisi, Nancy G. *Medieval and Early Renaissance Medicine: An Introduction to Knowledge and Practice.* Chicago, 1990.

Skeat (1872) — See Chaucer (1872).

Skelton *et al.* (1965) — Skelton, R. A.; Marston, Thomas E.; and Painter, George D. *The Vinland Map and the Tartar Relation.* New Haven, 1965.

Sotheby's (1952) — Sotheby's. Catalogue of sale on March 11, 1952. London, 1952.

Sotheby's (1961) — Sotheby's. *An Important Collection of Scientific Instruments (The First Part).* Catalogue of sale on Dec. 4, 1961. London, 1961. (Ref. 33, lot 136, plates 15, 16.)

Sotheby's (1968) — Sotheby's. *Important English and Continental Watches and Clocks and Scientific Instruments.* Catalogue of sale on Dec. 9, 1968. London, 1968. (Ref. 9-10, lots 21-23.)

Sotheby's (1986) — Sotheby's. *Fine Instruments of Science and Technology: 1300-1900*. Catalogue of sale on June 18, 1986. London, 1986. (Ref. 24, lot 125.)

Sotheby's S.A. (1985) — Sotheby's S.A. *Islamic Art.* Catalogue of sale on June 25, 1985. Geneva, 1985. (Ref. lot 236.)

Spitzer Catalogue (1892) — Catalogue of sale of the Frédéric Spitzer collection. Paris, 1892. (Ref. 2: 207, lot 2934.)

Stimson (1988) — Stimson, Alan. *The Mariner's Astrolabe: A Survey of Known, Surviving Sea Astrolabes.* Utrecht, 1988.

Stöffler (1513 and 1524) — Stöffler, Johann. *Elucidatio fabricae ususque astrolabii.* Oppenheim, 1513 and 1524.

Stöffler (1560) — Stöffler, Johann. *Traité de la composition et fabrique de l'astrolabe, & de son usage.... Le tout traduit du Latin de Iean Stofler de Iustingence....Avecques annotations... faites par Iean Pierre de Mesmes.* Paris, 1560.

Stone (1723) — Stone, Edmund. *The construction and principal uses of mathematical instruments, Translated from the French of M. Bion.* London, 1723.

Tanner (1587) — Tanner, Robert. *The Traveller's joy and felicitie, or a Mirror for Mathematics.* London, 1587.

Tardy (1972) — Tardy [pseud. of Henri Lengellé]. *Dictionnaire des horlogers français.* Paris, 1972.

Tavernari (1976) — Tavernari, Carla. "Manfredo Settala, collezionista e scienziato milanese del '600." *Annali dell'Istituto e Museo di Storia della Scienza di Firenze* 1, part 1 (1976), 43-61.

Taylor (1954) — Taylor, E. G. R. *The Mathematical Practitioners of Tudor and Stuart England.* Cambridge, England, 1954.

Taylor (1957) — Taylor, E. G. R. *The Haven-Finding Art.* New York, 1957.

Taylor (1966) — Taylor, E. G. R. *The Mathematical Practitioners of Hanoverian England 1714-1840.* Cambridge, England, 1966.

Thomas (1971) — Thomas, Keith. *Religion and the Decline of Magic.* New York, 1971.

Thompson (1977) — Thompson, Roger, ed. *Samuel Pepys' Penny Merriments.* New York, 1977.

Thorndike (1923-58) — Thorndike, Lynn. *A History of Magic and Experimental Science.* 8 vols. New York, 1923-58.

Thorndike (1944) — Thorndike, Lynn. *University Records and Life in the Middle Ages.* New York, 1944.

Thorndike (1949) — Thorndike, Lynn, ed. *The Sphere of Sacrobosco and Its Commentators.* Chicago, 1949.

Thorndike (1965) — Thorndike, Lynn. *Michael Scot.* London, 1965.

Times Atlas (1956/1959) — *Times Atlas of the World.* Vols. 2 and 4. London, 1956/1959.

Tomlinson (*c.* 1932) — *The John C. Tomlinson Collection of Portable Sundials.* [London], [*c.* 1932].

Turner, A. J. (1973) — Turner, A. J. "Mathematical Instruments and the Education of Gentlemen." *Annals of Science* 30 (1973): 51-88.

Turner, A. J. (1987) — Turner, A. J. *Early Scientific Instruments: Europe, 1400-1800.* London, 1987.

Turner, A. J. (1989) — Turner, A. J. "Paper, Print, and Mathematics: Philippe Danfrie and the Making of Mathematical Instruments in Late 16th Century Paris." In *Studies in the History of Scientific Instruments,* edited by Christine Blondel *et al.,* 22-42. London/Paris, 1989.

Turner, A. J. (in preparation) — Turner, A. J. "John Prujean." In *New Dictionary of National Biography.* Oxford, in preparation.

Turner, A. J., Catalogue (1974) — Turner, A. J. *Paper and Brass: Scientific Instruments and the Art of Printing. A Catalogue of an Exhibition held June...1974.* London, 1974.

Turner, A. J., Catalogue (1985) — Turner, A. J. *Astrolabes, Astrolabe Related Instruments.* The Time Museum: Catalogue of the Collection, edited by Bruce Chandler. Vol. 1: Time Measuring Instruments, part 1. Rockford, Ill., 1985.

Turner, G. L'E. (1986) — Turner, Gerard L'E. Paper presented at the meeting of the Scientific Instrument Commission of the IUHPS, Munich, 1986.

Turner, G. L'E. (1991) — Turner, Gerard L'E. "Navigazione." In *Storia della scienza: Gli strumenti,* 232-35. Turin, 1991.

Turner, G. L'E. (1994) — Turner, Gerard L'E. "The Three Astrolabes of Gerard Mercator." *Annals of Science* 51 (1994): 329-53.

Turner and Dekker (1993) — Turner, Gerard L'E., and Dekker, Elly. "An Astrolabe Attributed to Gerard Mercator, c. 1570." *Annals of Science* 50 (1993): 403-43.

Van Cittert (1954) — Van Cittert, P. H. *Astrolabes.* Leiden, 1954.

Van Cleempoel (1997) — Van Cleempoel, Koenraad. "The 'Philip II' Astrolabe." In *Power and Technology in the Sixteenth Century: Louvain as the Centre of Diffusion of Scientific Instruments.* Madrid, 1997.

Waterman (1997) — [Waterman, Trevor P.] *A Measure of Time: 25th Anniversary, Trevor, Philip, and Sons.* [London], [1997].

Waters (1966) — Waters, David W. *The Sea- or Mariner's Astrolabe.* Agrupamento de Estudos de Cartografia Antiga, no. 15. Coimbra, 1966.

Waters (1978) — Waters, David W. *The Art of Navigation in England in Elizabethan and Early Stuart Times.* 2nd ed. Greenwich, 1978.

Waters (1980) — Waters, David W. *Science and the Techniques of Navigation in the Renaissance.* 2nd ed. Greenwich, 1980.

Webster (1984) — Webster, R. S. *The Astrolabe: Some Notes on Its History, Construction and Use.* 2nd ed. Lake Bluff, Ill., 1984.

Webster and Webster (in preparation) — Webster, R. S., and Webster, M. K. *An Index of Western Scientific Instrument Makers to 1850.* In preparation.

Wede (1960) — Wede, Karl F. *Ship Models, Marine Antiques and Rare Maritime Books.* Saugerties, N.Y., 1960.

Wede (n.d.) — Wede, Karl F. *The Age of Sail: Ship Models for the Connoisseur.* Saugerties, N.Y., n.d.

Welborn (1931) — Welborn, Mary Catherine. "Lotharingia as a Center of Arabic and Scientific Influence in the Eleventh Century." *Isis* 16 (1931): 188-99.

White (1975)	White, Lynn, Jr. "Medical Astrologers and Late Medieval Technology." *Viator* 6 (1975): 295-308. Reprinted in White (1978), 297-315.
White (1978)	White, Lynn, Jr. *Medieval Religion and Technology: Collected Essays.* Berkeley, 1978.
Wiet Catalogue (1935)	Wiet, Gaston. "L'epigraphie arabe de l'exposition d'art persan du Caire." L'Institut d'Egypt. Cairo, 1935.
Yonge (1968)	Yonge, Ena L. *A Catalogue of Early Globes Made Prior to 1850 and Conserved in the United States.* American Geographical Society Library Series, no. 6. New York, 1968.
Yule (1903)	Yule, Henry, trans. and ed. *The Book of Ser Marco Polo The Venetian Concerning the Kingdoms and Marvels of the East.* 3rd ed., revised by Henri Cordier. 2 vols. London, 1903.
Zinner (1965)	Zinner, Ernst. *Astronomische Instrumente des 11 bis 18 Jahrhunderts.* 2nd ed. Munich, 1965.

Index

'Abbāsid Caliphs, 3
Abelard, Peter (1079-1144?), 7
Adelard of Bath (fl. 1116-1142), 10n
Adler, Max (1866-1952), ix, xi
ibn Adreth, Solomon, 23
Albertos, F. T., 65n
Alcock, Thomas (17th cent.), 163
Alexander VI, Pope (1431-1503), 7
Alfani, Alfano (late 15th cent.), 115, 116
Alfonso X of Castile (1221-1284), 36
Allen, John 1 (fl. 1631-1637), 162
analemma, 136
Anderton, Nathaniel (17th cent.), 163
Andulusia, introduction of astrolabe to, 6
  invention of universal astrolabe in, 88
Apianus, Petrus (1495-1552), 39, 85, 88, 130
De architectura (Vitruvius), 2
Arsenius, Gualterus (fl. 1554-1579), x, 8, 18, 37, 162
astrolabes by (#8, #9), 61-69
Asaro, Frank, 73
astrolabe clocks, 8
astrolabe-quadrants
  catalogued entries, by maker (if known)
    Harris, Daniel (#42), 143
    Hayes, Walter (#39), 136-38
    Prujean, John (#40), 139-40
    Sutton, Henry (#38), 134-35
    Thompson, Anthony (#37), 133
    Worgan, John (#41), 141-42
  catalogued entries, by place of origin
    England (#43), 144
    London (#37, #38, #39, #41, #42, #44, #45), 133, 134, 136, 141, 143, 145, 146
    Louvain (#36), 130
    Oxford (#40), 139
astrolabe-quadrants *(continued)*
  catalogued entries, by type
    Gunter (#37, #40, #42, #43, #44, #45), 133, 139, 141, 143, 144, 145, 146
    Panorganon or universal instrument (#39), 136
    Sutton (#38), 134
    universal (#36), 130
  Gunter-type, construction and parts of, 127-29
  of paper, 134
astrolabes
  astrological uses of, 12-14, 34
  in astronomical compendia, 8
  astronomical uses of, 3n, 10-12, 29
  catalogued entries, by maker (if known)
    Arsenius, Gualterus (#8, #9), 61-69
    Bos, Johannes (#14), 80-81, 117
    Danfrie, Philippe (#19), 23n, 95-97
    Fusoris, Johanne (Jean) (#2, #3), 44-48, 82
    Greene, Norman (#33, #34), 122
    Habrecht, Isaac 2 (#22, #23), 102-5
    Hartmann, Georg (#5, #6), 53-57
    Lucas-Dean, Rob (#31), 120
    Martinot, Ludovicus (#15), 23n, 82-84, 103
    Moreau, Jehan (Jean) (#19), 23n, 95-97
    Puig Aguilar, Roser (#32), 121
    Reeves, Edward Ayearst (#30), 118-19
    Schreckenfuchs, Laurentius (#12), 77-78
    Sevin, Pierre (#25), 109-10
    Vibrandi (#29), 117
    Zabeus, Bernardinus (#11), 75-76
astrolabes *(continued)*
  catalogued entries, by place of origin
    Barcelona (#32), 121
    Berkeley, Calif. (#33, #34), 122
    Ecton Brook, England (#31), 120
    England (#1, #16, #35), 40, 85, 123
    Europe (#7), 58
    Germany (#13, #17, #18, #20, #21, #26), 79, 90, 93, 98, 100, 111
    Italy (#14, #24), 80, 106
    London (#30), 118
    Louvain (#8, #9, #10), 61, 66, 70
    Memmingen, Germany (#12), 77
    Nuremberg (#5, #6), 53, 56
    Padua (#11), 75
    Paris (#2, #3, #19, #25), 44, 46, 95, 109
    Sens, France (#15), 82
    Spain (#4), 49
    Strasbourg, France (#22, #23), 102, 105
  catalogued entries, by type or projection
    Azarquiel (#32), 121
    Blagrave (#16), 85
    classic (#1-#7, #11, #12, #14, #15, #18-#20, #24, #28, #29, #31, #33-#35), 40, 44, 46, 49, 53, 56, 58, 75, 77, 80, 82, 93, 95, 98, 106, 114, 117, 120, 122, 123
    double planispheric (#22), 102
    la Hire (#30), 118
    multiple (#8, #9, #10, #21), 61, 66, 70, 100
    partial (#23), 105
    de Rojas (#17, #25, #26, #27), 90, 109, 111, 113
    surveyor's (#13, #24), 79, 108
  in Christian culture, 14, 23

astrolabes *(continued)*
classic-type, construction and parts of
alidade, 30, 31, 35
backplate, 30-31, 35
calendars, 31, 35
horse, 35
limb (rim), 30, 34
mater, 30, 34
rete, 30, 33, 35
ring and shackle, 30, 34
rule (regula), 33, 34
shadow square, 19, 35
tympan, 30, 33, 34
cultural diffusion of, 3-7
early treatises about, 3, 6, 59-60
history of, 2-24
in Islamic culture, 9-10, 13, 22, 23-24
in Jewish culture, 3n, 22-23, 59-60
mariner's, 15-18, 39, 148-49
as "mathematical jewels," 8
of paper, cardboard, or vellum, 8, 38, 77, 102, 105, 121
planispheric, 2, 36, 37
production and use in Europe, 6-23
topographical uses of, 15-21
universal, 36-39, 88
Astrolabium Catholicum, of Gemma Frisius, 37, 39
astronomical compendia, 8
Astronomicum caesareum (Apianus), 39, 88
Astronomy (Levi ben Gerson), 59-60
Atocha. See *Nuestra Señora de Atocha*
Aubrey, John (1626-1697), 13
Azarquiel. *See* al-Zarqāllu

Bar Ḥiyya, Abraham (fl. 12th cent.), 59
al-Battānī (d. 929), 10n, 19n, 59
Bion, Nicolas (*c.* 1652-1733), 8, 38
al-Bīrūnī, Abū Rayḥān (973-1048), 3, 13
Blaeu, Willem Janszoon (1571-1638), 8
Blagrave, John (*c.* 1558-1612), 8n, 10, 18, 38, 85, 89, 162
Blow, Edmund (fl. 1st half 18th cent.), 143, 162
Blundeville, William (fl. 1560-1602), 89
Bos, Jacob (2nd half 16th cent.), 80, 162
Bos, Johannes (fl. 1591-1623), 162
astrolabe by (#14), 80-81, 117
copy of astrolabe by (#29), 117
Brahe, Tycho (1546-1601), 10, 12
Brown, Thomas 1 (fl. 1627-1653), 163
Bylica, Martin (d. 1498), 21, 37

Catherine de Medici (1519-1589), 8n
Charles V of France (1337-1380), 7
Chaucer, Geoffrey (*c.* 1342-1400), 2, 13, 122, 123
China, introduction of astrolabe to, 4, 6, 13-14
Collins, John (1625-1684), 134, 135, 138
Cooke, Thomas 1 (mid-18th cent.), 143, 162

Danfrie, Philippe (*c.* 1532-1608), 8, 162
astrolabe by (#19), 23n, 95-97
Danti, Egnatio (1536-1586), 8, 116
Danti, Piervincenzo (2nd half 15th cent.), 114, 115, 116
Debeauvais, Francis, 104n, 105
Dee, John (1527-1608), 7
Delacerda, Don Luis (fl. mid-16th cent.), 65
Dominicus de Clavasio (fl. mid-14th cent.), 19
Dorn, Hans (*c.* 1435-1509), 21, 37
Drake, Stillman, xii
Dudley, Robert (1532/3-1588), 88

Eichner, Laurits Christian (1894-1967), mariner's astrolabe by (#47), 149
Elizabeth I of England (1533-1603), 7
Engelmann, Max (fl. 1st half 20th cent.), x
Europe, astrolabe production and use in, 6-23
ibn Ezra, Abraham (*c.* 1090-*c.* 1164), 23, 59

al-Fazārī, Moḥammad (fl. 760-790), 3
Foster, Samuel (fl. 1619-1652), 136
Frederick II, Holy Roman Emperor (1194-1250), 7
Fusoris, Johanne (Jean) (*c.* 1365-1436), 8, 131, 162
astrolabes by (#2, #3), 44-48, 82
Gamliel, Rabban II (fl. *c.* A.D. 80-116), 23
Gemma Frisius (1508-1555), 8, 10, 18, 21, 37, 39, 61, 162
George I of England (1660-1727), 163
Gerbert of Aurillac (Pope Sylvester II) (*c.* 945-1003), 6, 14, 19
Goldstein, Bernard, 58
Gosselin (17th cent.), 163
Greene, Norman, astrolabes by (#33, #34), 122
Gunter, Edmund (1581-1626), 127, 140
Gunther, Robert T., 115

Habrecht, Isaac 1 (1544-1620), 162
Habrecht, Isaac 2 (1589-1633), 162
astrolabes by (#22, #23), 102-5
Harington, John (late 16th cent.), 19n
Ḥarrān, early astrolabe manufacturing center, 3
Harris, Daniel (fl. 1735-1775), 162
astrolabe-quadrant by (#42), 143
Hartmann, Georg (1489-1564), 8, 162
astrolabes by (#5, #6), 53-57
copy of astrolabe by (#31), 120
Hayes, Walter (fl. 1642-1692), xii, 162
astrolabe-quadrant by (#39), 136-38
Heilbronner, Raoul (fl. 1890-1914), ix-xi
Héloise (*c.* 1098-1164), 7
Hermann Contractus (1013-1054), 6
Hipparchus (*c.* 150 B.C.), 2
Hire, Philippe de la (1640-1718), 38
hours, equal and unequal (planetary), 33, 34
Hūlāgū Khān (fl. mid-13th cent.), 4

India, introduction of astrolabe to, 3-4, 6n
Iraq, introduction of astrolabe to, 3

Jacquinot, Dominique (fl. mid-16th cent.), 8n
Jai Singh II (1686-1743), 4
Jamāl al-Dīn (fl. 2nd half 13th cent.), 4
Jean, Duc de Berry (1340-1416), 8, 8n, 9n, 162
John II of Portugal (1455-1495), 7

Kassell, Nancy, 78
Kepler, Johannes (1571-1630), 94
ibn Khalaf, 'Alī (fl. 11th cent.), 88
al-Khwārizmī (fl. *c.* 800-847), 10n, 19

King, David A., 88, 130
Köbel, Jacob (*c.* 1460-1533), 2
Krabbe, Johannes (fl. late 16th cent.), 8
Kublai Khān (1215-1294), 4, 13-14
Kynvyn, James (fl. 1569-1610), 89

Lahore, astrolabe manufacturing center, 3, 6n
Lattes, Bonet de (d. *c.* 1514), 7
Lauren, Christophor (fl. *c.* 1598), 163
Leo X, Pope (1475-1521), 7
Levi ben Gerson (1288-1344), 10, 12, 59-60
Leybourn, William (1626-1716), 136, 138
Liber particularis (Scot), 14
Libros del saber (Alfonso X of Castile), 36, 88
Lucas-Dean, Rob, astrolabe by (#31), 120

Maddison, F. R., 104n, 116
al-Ma'mūn (787-827), 3
Mandeville's Travels (Seymour), 14, 15
Manuel I of Portugal (1469-1521), 7
mariner's astrolabes, 15-18, 39
  catalogued entries (#46, #47), 148-49
Martinot, Ludovicus (fl. 1598-1631), 163
  astrolabe by (#15), 23n, 82-84, 103
Māshā'allāh (fl. *c.* 762-815), 2n
The Mathematical Jewel (Blagrave), 38, 85
Mensing, Anton W. M. (1st half 20th cent.), ix-xi
Mercator, Gerard (1512-1594), 61
Mercier (fl. 1st half 19th cent.), 46-47
Meteoroscopion, of Petrus Apianus, 39, 85, 130
Michel, Helen, 72-73
Moreau, Jehan (Jean) (fl. 1622-1628), 8, 10n, 163
  astrolabe by (#19), 23n, 95-97
Morrison-Low, A. D., 47n
Mörzer Bruyns, Willem, xn
Moskowitz, Saul, xii

Nebenzahl, Kenneth, xii
Norberg, Arthur, 73
North Africa, introduction of astrolabe to, 6
Nuestra Señora de Atocha, 148

Painswick astrolabe, 122
Palmer, John (fl. mid-17th cent.), 89
Panorganon: A Universall Instrument (Leybourn), 138
Pepys, Samuel (1633-1703), 163
Persia, introduction of astrolabe to, 3
Philoponus, John (fl. A.D. 530), 3
Planiglobium coeleste, et terrestre (Habrecht), 103
Planisphaerium (Ptolemy), 2
Polo, Marco (1254-1324), 4
Price, Derek J. de Solla, xii
Profatius quadrant, 127
Prophatius. *See* ibn Tibbon, Jacob ben Machir
projections
  Azarquiel or Gemma Frisius, 37
  Gunter, 128-29
  la Hire, 38
  orthographic, 37
  de Rojas, 37
  stereographic, 2, 29-30, 37, 127, 129
  Stöffler, 134
Prujean, John (fl. 1670-1706), 8, 163
  astrolabe-quadrant by (#40), 139-40
Ptolemy (fl. A.D. 150), 2-3, 10n
Puig Aguilar, Roser, astrolabe by (#32), 121

qibla, 24
quadratum nauticum, 18, 19n

ready reckoner, 91, 109, 111, 113, 118
Reeves, Edward Ayearst (1862-1945), astrolabe by (#30), 118-19
Ribero, Diego (fl. 1st half 16th cent.), 18
Rohde, Alfred, 115
de Rojas, Joanne (Juan) (fl. 1550), 37

Sacrobosco, John of (fl. 1230-1255), 15
Schissler, Christopher (d. 1609), xii, 9n, 88
Schreckenfuchs, Erasmus Oswald (1511-1579), 78, 163
Schreckenfuchs, Laurentius (fl. *c.* 1567), 163
  astrolabe by (#12), 77-78
Scot, Michael (1175?-1235?), 7, 14
Sebokht, Severus (pre-660), 3
The Sector on a Quadrant (Collins), 134, 138
Settle, Thomas, 116

Sevin, Pierre (fl. 1662-1688), 163
  astrolabe by (#25), 109-10
Sharp, Abraham (1651-1742), 88
Sheahan, D. B., xi, 137
Spain, introduction of astrolabe to, 6
  invention of universal astrolabe in, 88
  and medieval Hebrew astronomical tradition, 59
Stöffler, Johann (1452-1531), 18, 71, 134
Sutton, Henry (fl. 1648-1669), 8, 163
  astrolabe-quadrant by (#38), 134-35
Sylvester II, Pope. *See* Gerbert of Aurillac

Taqī al-Dīn (1525/6-1585), 10
Theon of Alexandria (fl. 2nd half 4th cent.), 3
Thompson, Anthony (fl. 1638-1665), 163
  astrolabe-quadrant by (#37), 133
Thompson, John, 163
ibn Tibbon, Jacob ben Machir (Prophatius) (*c.* 1236-1305), 127
Toledo, 37, 88
Tomlinson, John, xi
The Travels of Sir John Mandeville (Krása), 4
Turner, Gerard, 69n, 88

L'usage des astrolabes (Bion), 38

Vibrandi (*c.* 1900), 163
  astrolabe by (#29), 117
Vincent, Clare, 115
Vitruvius (d. post-A.D. 27), 2

Walcher (d. 1135), 6, 10, 10n
Webster, Roderick and Marjorie, xii
Wheatland, David Pingree, xii
Whitwell, Charles (fl. 1590-1611), 88, 89
Worgan, John (fl. 1682-1714), 163
  astrolabe-quadrant by (#41), 141-42

Zabeus, Bernardinus (fl. 1552-1559), 163
  astrolabe by (#11), 75-76
Zacuto, Abraham (*c.* 1450-*c.* 1522), 7, 18-19
al-Zarqāllu (Azarquiel) (d. 1100), 10n, 37, 38, 59